Aerospace Actuators 1

Series Editor
Jean-Paul Bourrières

Aerospace Actuators 1

Needs, Reliability and Hydraulic Power Solutions

Jean-Charles Maré

iSTE

WILEY

First published 2016 in Great Britain and the United States by ISTE Ltd and John Wiley & Sons, Inc.

ISTE Ltd
27-37 St George's Road
London SW19 4EU
UK

www.iste.co.uk

John Wiley & Sons, Inc.
111 River Street
Hoboken, NJ 07030
USA

www.wiley.com

Library of Congress Control Number: 2016936925

British Library Cataloguing-in-Publication Data
A CIP record for this book is available from the British Library
ISBN 978-1-84821-941-0

Contents

Introduction

This book is the first of three volumes that cover the topic of aerospace actuators following a system-based approach. The first volume provides general information on actuators and their reliability, and focuses on hydraulically supplied actuators. The second volume addresses more electrical actuators (electro-hydraulic, electro-hydrostatic and electromechanical) and their signal flows (Signal-by-Wire) as well as their power flows (Power-by-Wire). The third volume illustrates the concepts introduced in the two previous volumes by showcasing various examples of applications of actuation in aerospace (flight controls, landing gear and engines) as well as different types of aircraft (commercial, military or business aircraft, helicopters, convertibles and space launchers).

In order to successfully develop a system perspective in this book series, a top-down approach (from the requirements to the solution) has to be carefully combined with a bottom-up approach (from the technological maturity to the solution). The main guiding idea is, therefore, to focus on requirements and architecture (functional then conceptual) while revealing restrictions imposed by technology on concept implementation, in particular, with regard to functions and phenomena induced by technological choices. Indeed, in practice, solutions are designed according to the functional aspects that have to be implemented and combined. Subsequently, performance is assessed while taking into consideration technological shortcomings[1] such as magnetic hysteresis, winding inductance, mechanical

1 Depending on the application, it is possible for the same physical phenomenon in a technological component to either be harnessed to perform a specific function or to appear as a technological defect that threatens functional performance.

clearance or even hydraulic resistance of pipes. Architectural aspects are generally poorly documented in bibliographies compared to design and technological aspects. However, this does not mean that design and technology are of little importance. Quite the contrary, they too are essential aspects. Indeed, it very often turns out that, as of a given date, technology becomes the limiting factor for the industrial relevance of a project in terms of performance, reliability and cost. Similarly, the accuracy of mathematical models and the computing power they require often restrict safety margin improvement and design optimization.

Design, modeling and technological aspects are generally well documented in bibliographies but the architectural feature, both functional and conceptual, is usually rarely addressed. The difficulty, therefore, lies in formalizing architectures and concepts. The use of mathematical models and technology description is only resorted to when they are necessary for comprehension and technology selection in this volume. The "architecture"-oriented system approach highlights the required functionalities and interdependencies between system components. This approach shows more promise for improving global performance than the "component"-oriented approach. This is because a globally optimal system is hardly ever obtained by simply combining components that are each individually optimal. In commercial aeronautics, developing such an approach is made even more difficult by the organization of design offices: they are often divided up according to aircraft partitioning, and, therefore, often according to "trades". This partitioning was standardized for the first time by the Air Transport Association more than 50 years ago [A4A 14]. Nevertheless, a cross-cutting view has been increasingly encouraged by setting up plateaus, by promoting system expert positions and by the emergence of initial training in systems engineering covering both power and signal aspects.

In this first volume, emphasis is put on hydraulic power actuators. This choice can come as a surprise in this day and age where "more" electrical solutions and even "all" electrical solutions are almost always put forward. However, this choice is justified by several important considerations:

– Hydraulic technology is used extensively for actuating purposes on all aircraft, including the newer ones. The lifespan of a commercial aircraft is typically about 30 years and its marketing life also frequently extends beyond 30 years. New aircraft models of the 2010s should therefore typically still be flying in 2070.

– The maturation of more or all electrical solutions assessed in the laboratory can take more than a decade before reaching a stage satisfactory enough that the new technology can be implemented on aircraft. For instance, the electro-hydrostatic actuators that were implemented on Airbus A380 in 2007 had begun being developed in the mid-1980s.

– It is important to think in terms of requirements and performance rather than restrict choices to a given technological solution. In this regard, a more or all electrical solution should not be an end in itself; it should only be a means of providing safer, greener, cheaper and faster[2] services.

The progressive or complete removal of hydraulics teaching in engineering degrees contributes to a loss of initial skills for engineers. These two established facts advocate for a capitalization effort of knowledge that is prone to disappear.

Following the emphasis put on the system approach, requirements and architectures, this book develops an approach that is complementary to other existing publications on the topic. These constitute a significant source of information. Consequently, the following books should be recommended to broaden the scope of this series:

– generally speaking, for all aircraft systems, [CRA 99, DAN 15, MOI 01, ROS 00, SAU 09, WIL 01] and [WIL 08];

– for all aerospace actuating purposes, [RAY 93] and [SCH 98];

– for aerospace hydraulics, [GRE 85, JEP 85] and [NEE 91];

– for general hydraulics, [BLA 60, FAI 81, FAY 91, GUI 92, MAR 80, MER 67] and [VIE 80].

The present volume consists of seven chapters. Chapter 1 provides an overview of aerospace actuators with an emphasis on requirements and applications. Chapter 2 addresses reliability, that can heavily impact architecture choices in very early development stages. The following chapters focus on actuators supplied by hydraulic power sources. These chapters successively review the energy carrier function, the hydraulic fluid as well as its conditioning, power conversion, control and management; and finally actuators and hydraulic systems integration.

2 Concerning commercial aeronautics, it seems that the fast aspect is no longer one of the main goals because given the current state of technology, this aspect heavily impacts the other three (safety, environmental friendliness and cost).

Notations and Acronyms

Notations

Symbols

a	Weibull parameter	–
A	Availability	–
b	Weibull parameter	–
B	Bulk modulus	Pa
c	Speed of sound	m/s
C	Torque/Control port	N m/-
C_h	Hydraulic capacitance	m^3/Pa
C_p	Specific heat at constant pressure	J/kg/°C
C_q	Flow coefficient	–
d	Diameter	m
E	Energy	J
F	Force	N
f	Friction factor	–
$f(u)$	Density of probability	–
$F(u)$	Probability of failure	–
g	Gravitational constant	m/s^2
$h(u)$	Failure rate	depends
I	Electrical current	A
I_h	Hydraulic inertia	kg/m^4

J	Moment of inertia	kg m^2
k	Gain	depends
K	Mechanical stiffness	N/m
l	Length	m
L	Inductance	H
M	Mass	kg
n	Number of elements	–
N	Number of items in service	–
p	Lead	m
$\mathcal{P}$	Power	W
P	Pressure	Pa
Q	Volume flow rate	m^3/s
$R(u)$	Probability of success (or, in short, reliability)	–
R	Electrical resistance	Ω
s	Orifice cross-sectional area	m^2
S	Area	m^2
t	Time	s
u	Service quantity	depends
U	Electrical voltage	V
v	Linear velocity	m/s
V	Volume	m^3
V_0	Displacement	m^3/rad
x	Position	m
y	Position/opening	m
z	Vertical position	m
Δ	Difference	–
ε	Control error	depends
η	Efficiency	–
λ	Constant failure rate	/FH
μ	Dynamic viscosity	Pl
θ	Angular position	rad

Θ	Temperature	°C
ρ	Specific density	kg/m^3
τ	Time constant	s
ω	Angular velocity	rad/s
ω	Angular frequency	rad/s
ξ	Pressure drop coefficient	–

Subscripts

0	Initial, reference, no-load
a	Apparent
b	By-pass
c	Load/kinetic
C	Coulomb
d	Valve/metering/drain/difference
DC	Detection–correction
e	Elastic
em	Electromagnetic
f	Leakage
g	Gravity
h	Hydraulic
hc	Hydro-kinetic
hm	Hydro-mechanical
hs	Hydro-static
l	Laminar or linear
m	Mechanical/mean
M	Maximum
n	Rated
o	Orifice
p	Position/pilot
P	Supply/parallel
q	Quadratic/thermal
r	Reflected/response/feedback

R	Return
s	Structure
S	Series
t	Turbulent
u	Controlled
v	Volumetric
VM	Majority voting
x	Position
∞	Asymptotic

Superscripts

$*$	Setpoint
$-$	Per mass unit

Acronyms

ACMP	Alternative Current Motor Pump
ADP	Air Driven Pump
CSMG	Constant Speed Motor Generator
EBHA	Electric Backup Hydraulic Actuator
EBMA	Electric Backup Mechanical Actuator
ECAM	Electronic Centralized Aircraft Monitor
EDP	Engine Driven Pump
EHA	Electro-Hydrostatic Actuator
EMA	Electro-Mechanical Actuator
EMP	Electro-Mechanical Pump
FbW	Fly-by-Wire
FH	Flight Hour
HMA	Hydro-Mechanical Actuator
HSA	Hydraulic Servo Actuator
HSMU	Hydraulic System Monitoring Unit
LEHGS	Local Electric Hydraulic Generation System

MEPU	Monofuel Emergency Power Unit
MLA	Maneuver Load Alleviation
MRHS	Main Rotor Hydraulic Subassembly
MTBF	Mean Time Between Failure
MTOW	Maximum Take-Off Weight
MTTF	Mean Time to Failure
MTTR	Mean Time to Repair
PbW	Power-by-Wire
PCU	Power Control Unit
PoD	Power-on-Demand
PTU	Power Transfer Unit
PWM	Pulse Width Modulation
SbW	Signal-by-Wire
SPGG	Solid Propellant Gas Generator
RAT	Ram Air Turbine
THS	Trim Horizontal Stabilizer
TVC	Thrust Vector Control

1

General Considerations

1.1. Power transmission in aircraft

1.1.1. *Needs and requirements for secondary power and power flows*

On an aircraft, a distinction is made between primary power, which is used to ensure lift and airborne movement, and secondary power, which is used to power systems (flight controls, avionics, landing gear, air conditioning, etc.). Although much less significant than primary power, secondary power is nevertheless non negligible, as shown in Table 1.1.

Actuation (flight controls and landing gear)	Instantaneous power: 50–350 kW
Cabin lighting	15 kW permanently
Galley	120–140 kW intermittently (warming oven)
	90 kW permanently (cooling)
In-flight entertainment	50–60 kW permanently
Cockpit avionics	16 kW
Cabin air conditioning	190–300 kW

Table 1.1. *Secondary power requirements for a large commercial aircraft [COM 05]*

Power is generally conveyed from sources to users by redundant networks in electrical, hydraulic and pneumatic form. For a typical 300 seat aircraft, these networks are estimated to respectively transmit a power of 230 kVA, 230 kW and 1.2 MW.

Figure 1.1 illustrates the complexity of secondary power networks for a single-aisle aircraft of the Airbus A230 type [LIS 09]. On this diagram, power flows from power generators situated on the inner ring, through distribution networks located around the intermediate ring, to power users gathered on the third ring. The outer ring depicts the surrounding air which is considered here as being equivalent to a thermal power source. Power flows are depicted by colored arrows whose colors indicate the nature of the power involved.

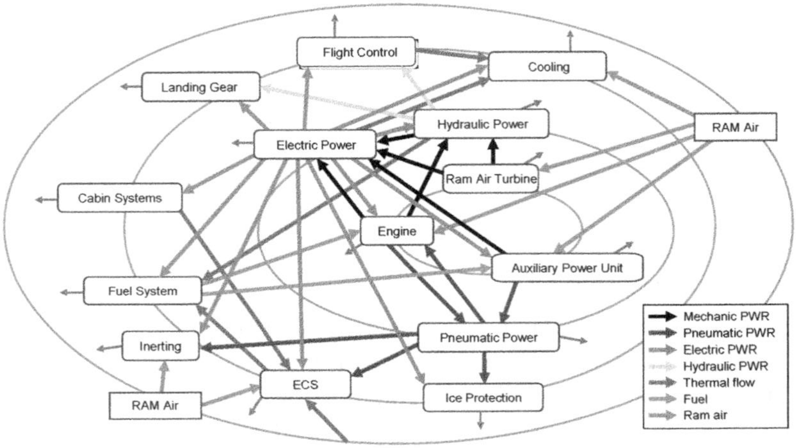

Figure 1.1. *Secondary power flows for an Airbus A230 type single-aisle aircraft [LIS 08]. For a color version of the figure, see www.iste.co.uk/mare/aerospace1.zip*

1.1.2. *Actuation functions*

A function can be defined as the act of transforming matter, energy or data in time, shape or space [MEI 98]. In practice, the perspective from which a function is viewed depends on the engineering task at hand. For instance, for the purpose of power scaling, the actuation function can be viewed as the transformation of power received at the source into power transmitted to the load; this transformation takes place both in shape (e.g. hydraulics toward translational mechanics) and in space (aspect of power transmission from point A to point B). In contrast, when designing flight controls, the actuation function is considered as the act of converting a signal (e.g. an electrical command for positioning a load) into another signal (current position of the load).

Power requirements for actuation are numerous and diverse. They essentially concern the following.

– *Primary flight controls*

The purpose of primary flight controls is control of aircraft trajectory. On a conventional aircraft, as the one pictured in Figure 1.2, they take the form of control surfaces responsible for controlling the three rotational degrees of freedom: the ailerons for roll, the rudder for yaw and the elevator for pitch.

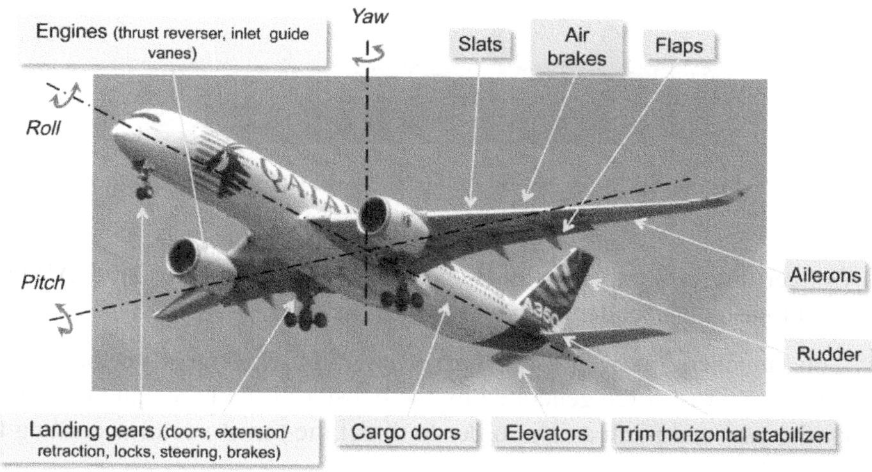

Figure 1.2. *Actuation needs on a commercial aircraft*

As for helicopters, they offer four degrees of freedom. Helicopter flights are controlled by acting:

– on the swashplate, Figure 1.3. The swashplate translation with respect to the rotor axis makes it possible to collectively act on the pitch of the blades in order to act on the intensity of the lift vector generated by the main rotor. Tilting the swashplate about the two axes perpendicular to the rotor axis causes the pitch to vary cyclically (pitch of the main rotor blades during one rotor revolution). In turn, this allows the rotor lift vector to be tilted about the roll or the pitch axis;

– on the collective pitch of tail rotor blades for yaw control.

Figure 1.3. *Swashplate actuation on an AS332 helicopter*

Convertible aircraft such as Boeing V22 or Agusta-Westland AW609 also put to use the nacelle tilt about the pitch axis.

– on launchers (as well as on fighter aircrafts with thrust vector control feature), the thrust force generated by the booster or the jet engine is steered about the yaw and pitch axis in order to direct the thrust vector according to the desired trajectory;

– mechanically signaled flight controls also rely on actuators to superimpose on the pilot's setpoint, the demands of the autopilot as well as stability and control augmentation commands.

– *Secondary flight controls*

Secondary flight controls make it possible to modify the aerodynamic configuration during particular flight phases. On conventional aircraft, slats and flaps increase the chord and curvature of the wings. This is done to increase the lift of wings at low speeds and therefore decrease the takeoff or landing speeds. Airbrakes (also called spoilers) reduce aircraft speed by increasing aerodynamic drag. Trim tabs, for instance the trimmable horizontal stabilizer, ensure the global equilibrium of the aircraft during the given rectilinear flight phases (e.g. climb, cruise or approach) so that primary flight controls operate around their neutral position on average.

– *Landing gears*

These require numerous actuation functions:

– for raising or lowering landing gear by sequencing the opening or closing of doors, extending or retracting the gear and locking it in a raised or lowered position;

– for steering the wheels in order to ensure steerability on the ground during taxiing;

– for wheel braking in order to dissipate as heat part of the kinetic energy associated with the horizontal speed of the aircraft (in addition to airbrakes and thrust reversers during landing). Left/right differential braking can also contribute to improve steerability on the ground;

– also worth mentioning, landing gear struts are autonomous hydropneumatic components. Upon touchdown, they absorb the kinetic energy associated with the vertical speed component of the aircraft with respect to the ground.

– *Engines*

Engines also rely on actuators to steer inlet guide vanes on the turbine stator, to deploy or stow thrust reversers, to operate maintenance panels, to modify the geometry of air intakes or nozzles, or even to control propeller blade pitch.

– *Utilities*

Other actuators are also used, for example, to operate cargo doors (and passenger doors on new large aircraft such as Airbus A380 and Boeing 787), rotor brakes for helicopters, winches, weapon systems (aiming guns, raising or lowering the arresting hook, etc.) among other things[1].

1.1.3. *Actuation needs and constraints*

The solutions implemented for actuation in aeronautics and space have to meet numerous requirements and comply with the following strict constraints.

1 The supersonic transport Concorde aircraft also used nose lowering actuators to give pilots an unblocked view of the runway.

– *Type of mission*

On the flight timescale, the need for actuation can be seen as continuous, such as, for example, the need for primary flight controls. However, this need can also be considered transient, meaning that it only exists during a minor portion of the mission. For example, this is the case for the landing gear steering function or for secondary flight controls. Lastly, the need for actuation is said to be impulsive when it only appears for a very short amount of time. An example of this is landing gear unlocking.

– *Controls*

The vast majority of actuators are closed-loop position controlled (e.g. for flight control surfaces or for steering nose landing gear). Although it is less frequent, they can also be closed-loop speed controlled (e.g. to drive back-up electric generator hydraulically) or even closed-loop force or pressure controlled (e.g. for braking). Additionally, some controls can be of the on/off type, such as, for example, landing steering locks.

– *Power and dynamics*

Summed up in Table 1.2 are examples of power and dynamics needs as a function of the type of aircraft.

Actuation function	Typical range	Aileron Airbus A320	Nose landing gear steering Airbus A320	Tiltrotor BoeingV22 OspreyMode conversion	Thrust vector control Ariane V
Stroke (mm) (degree)	20–700	44	 ±75	1143	±160
Speed (mm/s) (degree/s)	20–500 10–90	90 no-load	 20	97	972 no-load
Force (kN) (Nm)	20–350	44	 7000	80	347
Bandwidth (Hz)	1–20	≈ 1	≈ 1.5	3.2	7.9

Table 1.2. *Examples of power needs*

It is important to keep in mind that the power consumption of actuation functions indicated here corresponds to the worst case scenario. In reality, under normal circumstances, actuation functions are operated well below these extreme values. Figure 1.4 illustrates this statement for the actuator of an Airbus A320 aileron. It can clearly be seen that during the course of a typical mission, only −80 to 20% of the available force and ±15% of the available speed are used (except checklist).

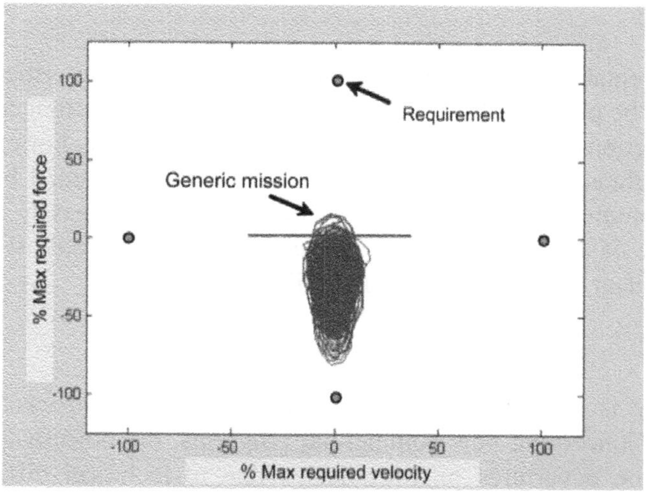

Figure 1.4. *Power requirement for the actuator of an Airbus A320 aileron [MAR 09]*

– *Environment*

Actuators are exposed to harsh climate conditions (pressure, temperature and humidity) as well as harsh electromagnetic (interference and lightning) and vibration environments. Every time they fly, actuators undergo a pressure and temperature cycle. For instance, at the cruising altitude of a jet (10,360 m or 34,000 ft), absolute pressure is only as high as 250 mbar and the temperature drops to −52.3°C (for a standard atmosphere [ICA 93]). Regarding hydraulics, the major constraint faced is maintaining the fluid temperature in its normal operating range. For example, actuators must be functional between −60°C and +100°C and they must be operational, meaning achieving full performance, between −40°C and +70°C.

– *Lifespan*

The lifespan of aircraft typically varies from 5,000 flight hours for fighter jets to more than 100,000 flight hours for new commercial aircraft (10,000 h for a NH90 helicopter, 48,000 h for the first Airbus A320 aircraft, 140,000 h for an Airbus A380). This lifespan generally corresponds to operating over the course of 30–40 years.

– *Reliability*

The acceptable probability of a failure depends on the criticality of the function to be performed (see Chapter 2). Since actuators often contribute to critical functions, tolerated failure rates are extremely low. For example, for primary flight controls, one catastrophic event is tolerated per 1 billion flight hours in commercial aeronautics. This major constraint heavily impacts the architecture of actuation systems. These systems therefore most often have to be redundant in order to respond to failure as required.

– *Maturity*

Maturity is a strong sales argument that directly impacts operational readiness. Concerning new commercial programs, the objective is to reach 99% on-time departures or with delays due to technical difficulties not exceeding 15 min.

– *Topology*

On aircraft, several dozen actuators are generally implemented and can be located dozens of meters away from their power source. Weight, position and performance of the power delivery network are therefore heavily impacted by the spatial layout of hydraulic systems. On an Airbus A380, there is, for example, more than 40 flight control actuators [MAR 04], some of them located more than 60 m away from the hydraulic power generator.

1.2. Primary and secondary power transmission functions for actuators

Generally speaking, the main functions associated with power transformation and metering[2] are clearly identifiable (see Chapters 4 and 5).

2 Power metering is also known as “control of power”.

Conversely, when dealing with actuation solutions, other significant functions are often neglected even though these functions turn out to be the most difficult to master in practice. It is therefore essential to pay close attention to:

– reversibility, a concept which makes it possible for a passive actuator, for example, to let itself be driven by an active actuator through the load they share;

– protection against excess static and dynamic force, which restricts mechanical stress/strain on control surfaces or on the airframe, for example, during sudden gusts of wind;

– cooling or heating in order to maintain actuator temperature within its normal operating or functional range;

– damping, to dissipate energy and avoid resonance. For instance, to avoid shimmy[3] of the nose landing gear steering;

– Dissipation of the energy to be absorbed when reaching the end-stop. For example, this is important for thrust reversers;

– force equalization, which is intended to ensure that actuators equally share the responsibility of driving a single load without force-fighting. For example, for the three active actuators of a single rudder;

– motion synchronization, which aims to position identically and at all times independent loads, each fitted with their own actuator. For example, for two independent panels of a single thrust reverser;

– locking in position, de-clutching or returning to a neutral configuration depending on the desired response to failure;

– maintenance or diagnosis. For example, with the purpose of isolating part of the system to detect and measure a possible internal leak.

In order to bring structure to this study of architectures, it is useful to define and distinguish the primary functions and secondary functions. A detailed investigation of these functions and the technological or conceptual architectures that enables their implementation will be provided in the following chapters regarding hydraulically powered actuators.

3 Sustained vibration of landing gear in rotation about the gear strut axis [CUR 88].

1.2.1. *Primary functions*

The general structure of the architecture of a power transmission system can be represented by the diagram in Figure 1.5 [CND 02]. Flows of information (signal component) are distinguished from flows of energy (power component) on the schematic.

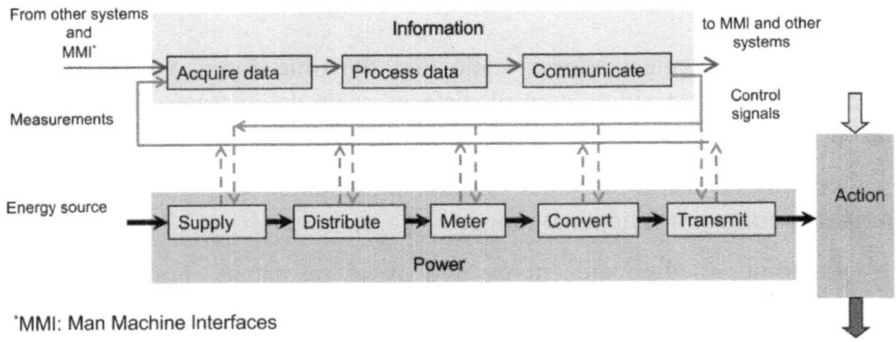

Figure 1.5. *Functions in the information and power chains [CND 02]*

According to this representation, power architecture involves five key functions: to supply, distribute, meter, convert and transmit. This interesting schematic point of view calls for a number of comments:

1) On the path between sources and users, generic functions of the power chain can appear several times and in a different order.

2) On this diagram, the signal architecture is explicitly separated from the power architecture. However, interface functions between the signal and the power components ("measure" and "apply command") are not mentioned.

3) The information chain can additionally include monitoring functions (usage, diagnosis and prognosis).

4) Within aerospace actuating components and systems, signal and power functions are often performed hydromechanically for the sake of simplicity, compactness and reliability. This partition of signal and power is often hard to see, all the more since it is seldom emphasized in the schematics of technical documents. This aspect will be noticeable in several examples over the course of the next chapters.

5) In order to perform its functions, the information chain also requires a power supply; for instance, to power flight control computers. Therefore, it can also be seen as a power chain if interested in this perspective: such as when the chain is used to assess the thermal equilibrium of electronic cards or to assess consumption in case of loss of the normal power supply. It is essential to keep this in mind while designing architectures, especially with respect to reliability requirements.

6) Misunderstandings often arise from the interpretation of the word actuator. Indeed, the meaning of this word is different from one community to the other. For instance, in the IEEE (Institute of Electrical and Electronics Engineers) community, an actuator often represents the component that converts power between electrical and mechanical domains, typically an electric motor. In the field of aeronautics, an actuator is instead loosely defined as the physical device that the aircraft manufacturer is provided with and integrates in the aircraft to carry out the actuation function. Depending on the level of integration, the "actuator" can include power metering, power transmission and even control functions in addition to its power converting function. For example, the actuator of an Airbus A350 aileron WXB introduced in Figure 1.6 incorporates both power and position control chains.

Figure 1.6. *Airbus A350 XWB aileron actuator incorporating position control electronics*

1.2.2. *Secondary functions*

In practice, the power section performs many secondary functions, the significance of which is far from minor. As a matter of fact, these secondary functions are the ones that cause the most specification, design and maintenance issues. They are essentially linked to safety and to power management depending on the mode of operation. Key secondary functions are as follows:

– *Concentrate.* Several power sources are combined to feed one or more users, while avoiding interactions between sources. For example, several pumps can feed a hydraulic power network.

– *Divide.* The power supply is divided between several users, depending on demand or following a given proportion. For example, a single network feeds several flight control actuators.

– *Isolate.* The component is isolated from the rest of the power circuit. For example, the system responsible for extending the landing gear does not need to be powered during cruise.

– *Restrict.* Onboard systems are protected from excessively high values, and sometimes also from excessively low values of variables associated with power transfer. For example, the hydraulic network pressure is limited in the event that the constant pressure regulation of hydraulic pumps fails.

– *Ensure irreversibility.* Power propagation is enabled in only one direction, generally from the source to the load. This makes it possible to freeze the trimmable horizontal stabilizer in position in the event that the actuation fails, for example.

– *Prioritize power supply.* In case of low power, power supply to non-essential users is cut. For example, primary flight controls are prioritized over landing gear extension that, as a consequence, is left to extend under its own weight.

– *Store and restore energy.* Energy is stored and restored according to demand and mode of operation. For example, to maintain the parking brake active after engines are switched off.

– *Condition hydraulic fluid.* Hydraulic fluid is maintained in good condition to ensure it is able to properly carry out all required functions (power transmission, lubrication and heat transfer).

1.2.3. *Signal approach and power approach*

When a signal approach is adopted, it is implicitly assumed that the element on the receiving end of the signal has no impact whatsoever on the signal it receives. This approach is interesting from a functional point of view. However, it soon starts showing flaws when considering power transmission systems. Indeed, no matter the physical domain, power that flows along a functional link is the product of two power variables. And these variables are linked by elements located at the source and at the end destination of the power link. Table 1.3 summarizes power variables for electric, mechanic and hydraulic domains. Useful energy variables and units are also indicated in the table.

	Power variables		Useful energy variables
	Effort variable	Flux variable	Flux integral with respect to time
Electrical	Voltage (V)	Current (A)	Electric charge (C)
Mechanical			
Translation	Force (N)	Linear velocity (m/s)	Linear position (m)
Rotation	Torque or moment (Nm)	Angular velocity (rad/s)	Angular position (rad)
Hydraulic	Pressure (Pa)	Volume flow rate (m^3/s)	Volume (m^3)

Table 1.3. *Power and energy variables*

On architecture schematics, it would be interesting to explicitly indicate which of the two approaches, signal or power, has been adopted by using a different graphical representation for each. The two approaches could, for example, be differentiated by making lines representing power flows thicker compared to information flows lines. Furthermore, the physical domain concerned could be highlighted by associating it with a specific color.

1.2.4. *Types of actuators*

Denomination of aerospace actuators is determined both by the technological domain (mechanical, hydraulic or electrical) used for interfacing them on the signal level (controls and measurements) and used for their power supply. Regarding solutions powered electrically, the technological domains (hydraulic or simply mechanical) used within the actuator to convert electrical power into mechanical power have to be even further specified. Table 1.4 summarizes the main types of actuators that are encountered in aircraft. Uppercase letters M, H and E represent the different types of technologies: M for mechanical, H for hydraulic and E for electrical.

Signal Interface	Interface and internal power conversion	Most common denomination
M**	M*	Manual
M**	H → M	Hydromechanical (HMA stands for Hydro Mechanical Actuator)
E***	H → M	Servo-hydraulic (HSA stands for Hydraulic Servo Actuator)
E	E → H → M	Electro-hydrostatic (EHA stands for Electro Hydrostatic Actuator)
E	H → M (normal) E → H → M (backup)	Servo-hydraulic with electro-hydrostatic backup (EBHA stands for Electric Backup Hydraulic Actuator)
E	E → M	Electromechanical (EMA stands for Electro-Mechanical Actuator)
E	H → M (normal) E → M (backup)	Hydraulic with electromechanical backup (EBMA stands for Electrical Backup Mechanical Actuator)

* Optionally control tabs can provide assistance by aerodynamic force to flight controls
** Optionally with additional control signals generated hydraulically or electrically (autopilot, stabilization, etc.)

*** Optionally with mechanical backup

Table 1.4. *Types of actuators according to the nature of their signal and power interfaces*

The following paragraphs will further describe the actuators introduced in Table 1.4.

– *Manual actuation*

When actuation is carried out manually, it can be aerodynamically assisted for power, by using control tabs for example [ROS 00]. Furthermore, manual actuation can be assisted from an information standpoint by secondary actuators, such as when performing an autopilot (Figure 1.8) or stabilization function.

– *Hydromechanical actuators (HMA)*

Hydromechanical actuators (HMA) have been the "go-to" actuators for aircraft up until the emergence of electrical controls, such as those chosen for Airbus A300-600 spoilers. With very few recent exceptions, flight controls of helicopters still use this type of actuator.

– *Hydraulic servo actuators (HSA)*

The emergence of computerized electrical controls was made possible by the development of hydraulic servo actuators (HSA). Indeed, within these actuators, a servovalve serves as a power interface between electrical and hydraulic domains. Airbus A320 was the first ever commercial aircraft to be fitted with HSA and it initiated the universalization of these new actuators for flight controls. However, at the time, these servo actuators were the first of their kind and therefore still retained a mechanical backup for the information chain of certain actuators, such as the mechanical backup that was entirely removed from Airbus A380. Electrically controlled actuators have become widespread for flight controls as well as for braking and landing gear steering.

– *Electro-hydrostatic actuators (EHA)*

Over the past decade, electro-hydrostatic actuators (EHA), that are powered electrically, have appeared. EHAs transmit power to the load through a hydrostatic loop which involves, within the actuator, an electric motor, a pump and a hydraulic cylinder. These solutions are used as backup actuators for the time being, such as on Airbus A380, A400M and A350 XWB aircraft.

– *Electro-hydrostatic backup servo-hydraulic actuators (EBHA)*

Airbus A380 also introduced electro-hydrostatic backup servo-hydraulic actuators (EBHA) that operate as servo-hydraulic actuators in regular mode and as electro-hydrostatic actuators in backup mode. As a consequence, they are supplied both by hydraulic power and by electrical power.

– *Electromechanical actuators (EMA)*

Electromechanical actuators (EMA) combine an electrical motor with mechanical transmission. Their use is still not very widespread when it comes to performing high power critical functions due to the lack of maturity of certain of their functions. Boeing B787 relies on EMAs for 4 out of 14 spoilers and for wheel braking.

Technological solutions for power transmission within the actuator can be combined in different ways [MAR 11]. For example, in order to implement a backup mode dissimilar from the normal mode, in Airbus A400M, landing gear doors are driven by electromechanical backup hydraulic actuators (EBMA).

This first volume is focused on mechanical power transmission using hydraulic technologies. Most or all electrical solutions, both on the signal level and on the power level, will be addressed in the second volume of this series. Mechanical to mechanical power conversion – even though it can be found in certain actuation functions powered hydraulically – will also be addressed in the second volume.

1.3. Hydraulic power actuation

The following section provides a quick introduction and overview of hydraulically powered actuation, which will then be further detailed in the next chapters.

1.3.1. *Units and reference values*

Table 1.5 summarizes the symbols and units used in power hydraulics. As is very often the case, international systems of units (SI units) have the benefit of constituting a homogeneous unit system. However, they are ill-suited for quick manual calculations and they hamper the mental representation of physical quantities. "Engineering" units, which include metric (excluding SI units) and imperial units, make up for this disadvantage by scaling down SI values of physical quantities "to a human level", meaning these scaled values typically vary from 0.1 to 1,000. However, such units should be used with care because they do not constitute a homogeneous unit system for calculations. Therefore, if not used to deal with such units, it is best to carry out calculations using SI units and then to convert results into "engineering" units (metric or imperial) to assess their plausibility and communicate with experts on these results.

	Pressure	Volume flow rate
Symbol	P	Q
SI Units	Pascal (Pa or N/m^2)	m^3/s
Metric Units (excluding SI units)	bar (1 bar = 10^5 Pa)	l/min (1 l/min = 1/60,000 m^3/s)
Imperial Units	pound/inch^2 (psi) 1,000 psi = 69 bar	inch^3/s (cis) 1 cis = 0.983/min gal/min (gpm) 1 gpm = 3.785 l/min

Table 1.5. *International and common units used in the field of hydraulics*

If the fluid is assumed to be incompressible and of constant temperature, then power $\mathcal{P}$ supplied by the fluid between two points 1 and 2 of a circuit is given by:

$$\mathcal{P}=Q(P_1-P_2) \qquad [1.1]$$

A quick calculation can be carried out using "trade" units by introducing a homogeneity factor of 600:

$$\mathcal{P}_{\text{kW}}=Q_{\text{l/mn}}(P_1-P_2)_{\text{bar}}/600 \qquad [1.2]$$

1.3.2. *Energy transport by a liquid*

There essentially exists four ways of transporting energy using an incompressible fluid: by varying its altitude, its velocity, its pressure or its temperature.

– *Gravity transport*

When raising the altitude of an incompressible liquid by a quantity z (m), its gravitational specific energy $\overline{E}_g$ (J/kg) is increased by:

$$\overline{E}_g = gz \qquad [1.3]$$

where g (m^2/s) is the acceleration due to gravity.

For example, for an increase in altitude of 5 m on Earth where $g = 9.81$ m/s², the gravitational potential energy of 1 kg of fluid increases by 49 J. This solution is widely applied in hydroelectric power plants. However, it is not applicable in most other applications because the variation in altitude is relatively weak and is enforced by geometric considerations. The effect of gravity is therefore considered as a disturbance. Indeed, between two points of the same hydraulic line spaced apart by an altitude difference z, the gravity effect induces a pressure difference:

$$\Delta P_g = \rho g z \qquad [1.4]$$

where ρ (kg/m^3) is the specific density of the fluid.

For example, on the same return line of a commercial airplane, the pressure at a rudder actuator is typically lower by 1 bar than the pressurized reservoir pressure, assuming the latter is located 10 m lower (ρ= 1,020 kg/m³).

– *Transport in (hydro)kinetic form*

When increasing the velocity of an incompressible liquid from v_1 to v_2 (m/s), its hydrokinetic specific energy $\overline{E}_{hc}$ is increased by:

$$\overline{E}_{hc} = \frac{1}{2}(v_2^2 - v_1^2) \qquad [1.5]$$

For example, a fluid domain that goes from zero to 10 m/s acquires 50 J/kg of hydrokinetic specific energy.

Therefore, in order to be of interest, energy transmission in kinetic form has to involve high speeds. Unfortunately, high speeds generate significant pressure drops in lines and give rise to loud noise and significant dynamic phenomena[4]. As a consequence, energy transmission in kinetic form is hardly ever implemented. Furthermore, converting kinetic energy into pressure energy is a very inefficient process. This contributes to further minimizing the benefit of power transport in hydrokinetic form.

– *Transport in hydrostatic form*

When raising the pressure of an incompressible liquid by a quantity ΔP (Pa), its hydrostatic specific energy $\overline{E}_{hs}$ is increased by:

$$\overline{E}_{hs} = \frac{\Delta P}{\rho} \qquad [1.6]$$

Consider a standard pressure difference of 200 bar and a specific density of 1,000 kg/m^3, in these conditions 1 kg of liquid has a pressure energy equal to 20,000 J.

Therefore, compared to the two previous transport methods described, energy transport in hydrostatic form reduces the required fluid mass four hundred fold!

– *Transport in heat form*

When raising the temperature of an incompressible liquid by a quantity $\Delta\Theta$ (°C), its heat specific energy $\overline{E}_{\Theta}$ is increased by:

$$\overline{E}_{\Theta} = C_p \Delta\Theta \qquad [1.7]$$

where C_p is the heat capacity of the hydraulic fluid (J/kg/°C).

When 1 kg of liquid, with a specific heat equal to 1,900 J/kg/°C, undergoes a temperature increase of 10°C, its heat energy increases by 19,000 J.

---

4 For this reason, fluid velocity in lines is limited to a few meters per second.

For the purpose of actuation, the main goal is to transport energy using liquid in order to, in the end, make it possible to apply a sufficiently large force on mechanical loads at low speeds. This is why the following three concerns argue in favor of energy transport in hydrostatic form:

– as for kinetic energy, converting heat energy into hydrostatic energy is a very inefficient process. However, this is not the case for the reverse process;

– regarding positive displacement hydraulic-mechanical power transformers (cylinders, pumps and motors), force is functionally proportional to pressure and velocity is functionally proportional to volume flow rate. Therefore, it is "natural" to transport energy in pressure form since it is required to transmit large forces;

– the temperature difference possible to harness and exploit for energy transmission in heat form is extremely small when aiming to cover the entire operating range of the aircraft.

Table 1.6 gives an example to illustrate the benefit of transporting energy in hydrostatic form. The masses of fluid consumed by each means of energy transport are compared. This comparison is carried out under the assumptions that: an aircraft rudder has to be steered 1° and that this requires the application of an average moment of 800 Nm, which corresponds to an energy of 14 J. It is also assumed that conversions are 100% efficient.

	Hypothesis	Specific energy (J/kg)	Consumed fluid mass (g)	Comments
Gravity	$g = 9.81$ m/s^2 $z = 5$ m	49.05	285	Height imposed by geometric considerations
Hydro-kinetic	$v_1 = 0$ m/s $v_2 = 10$ m/s	50	280	Poor efficiency of the conversion process of velocity energy into pressure energy
Hydro-static	$\Delta P = 200$ bar $\rho = 1000$ kg/m^3	20,000	0.7	Solution implemented Excellent efficiency of the conversion process of pressure energy into kinetic energy
Heat	$\Delta\Theta = 10$°C $C_p = 1900$ J/kg/°C	19,000	0.738	Very poor efficiency of the conversion process of heat into work

Table 1.6. *Benefit of transporting energy in hydrostatic form*

Even without taking into account the efficiency of conversion processes, the benefits of transporting energy in hydrostatic form are obvious. Since the mass of fluid required for transporting and converting power is directly proportional to the differential pressure, it is clearly beneficial to elevate working pressures, at least from the point of view of the fluid. In practice, this pressure raise is limited by several effects. First, the thickness of sleeve casings (lines, pipes, cylinders, etc.) must be increased in order to resist higher mechanical stresses. Second, a raise in pressure – which allows for lines with smaller diameters – increases fluid inertia which, in turn, increases the amplitude of the dynamic phenomena it gives rise to. Finally, a pressure raise requires a great command of sealing and lubrication technologies in order to ensure the lifespan and reliability required. At any given time, the acceptable working pressure is defined by the balance of all these considerations.

REMARKS.–

– From the previous equations it is possible to carry out simple calculations to obtain orders of magnitude. If converting hydrostatic energy into hydro-kinetic energy is performed with 100% efficiency, free flow from a pipe at 200 bar to the open of an incompressible liquid of specific density 1,000 kg/m^3 generates a speed of 200 m/s. Furthermore, if the stream is vertical, the fluid rises up to 2,000 m high, given that the conversion of hydrostatic energy into hydro-kinetic energy is also perfect. Conversely, if all the pressure energy is dissipated as heat transmitted to the liquid flowing through a restriction, the temperature of the liquid increases by 10.5°C for a fluid with a heat capacity of 1,900 J/kg/°C.

– Since a liquid can never be rigorously incompressible, it is interesting to quantify the energy associated with its elastic deformation. For a pressure increase ΔP, a fluid with an effective Bulk modulus B stores, because of its deformation, a specific energy equal to:

$$\overline{E}_c = \frac{\Delta P^2}{2\rho B} \quad [1.8]$$

For instance, for an increase in pressure by 100 bar, the hydraulic liquid of a commercial aircraft (ρ= 1,000 kg/m^3 and B = 10,000 bar, in practice) stores a specific deformation energy of 50 J/kg and the relative variation of fluid volume is 1%.

– The four ways of transporting energy using a fluid explicitly appear when they are put together to form the energy equation.

1.3.3. *Historical evolution of power and pressure use*

Increase in power requirements

Hydraulic power implemented in aircraft has been consistently increasing since the dawn of aeronautics. As proven by the many publications on aeronautic hydraulics released between 1937 and 1950, the use of hydraulics first spread to assume transient or pulsed functions. Examples of such functions are shown in Figure 1.7 and they include: extending and retracting landing gear, braking, drawing wing-flaps in and out or even controlling engine cooling cowl flaps.

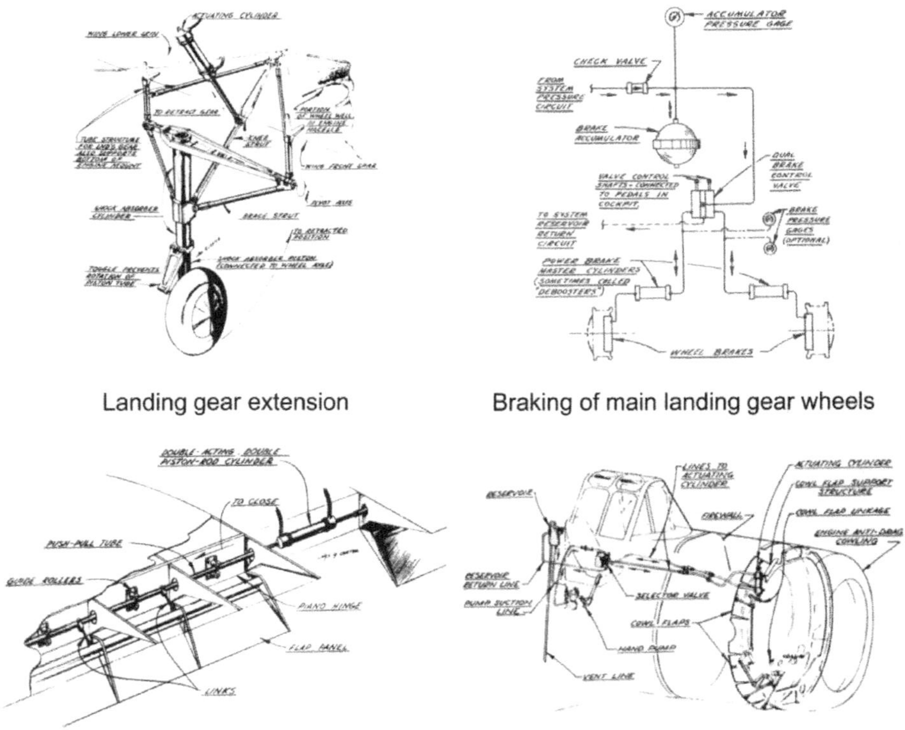

Landing gear extension | Braking of main landing gear wheels

Control of high-lift devices (flaps) | Control of engine cooling cowl flaps

Figure 1.7. *First applications of hydraulics to aerospace [THO 42]*

Furthermore, in very early days, hydraulics was also implemented to assume continuous autopilot functions. Their purpose was to move control surfaces of primary flight controls by acting on cables or on control linkages in order to maintain a given trajectory in spite of aerodynamic disturbances. The concept of a "gyropilot", used since the 1920s on boats, was applied by Sperry to aircraft in its three-axis version in the late 1930s. For instance, it was equipped on the Lockheed L-14 as soon as 1937. Figure 1.8 illustrates its operating principle for the roll axis. The detection of undesired roll is performed by a mechanical gyroscope (A) whose rotor is put in rotation by the airflow suctioned from the gyroscope thanks to a vacuum pump (B). In the presence of roll, the gyroscope develops a control command via the rotation of the gimbal ring carrying the rotor. This rotation induces a differential depression on the control membrane (F) of a hydraulic valve (H). The opening of the hydraulic valve meters the amount of hydraulic power transmitted from the pump (J) to the hydraulic cylinder (K) which operates ailerons (M, M') through cable transmission (L, L'). In manual flight mode, the cylinder chambers are connected via the valve (Z) opening. The cylinder rod slides freely and does not resist to displacements initiated by the pilot. Hence, less than 35 years after the first motorized flight of the Wright brothers, aircraft were already equipped with three-axis autopilots that was purely mechanical, pneumatic and hydraulic.

During the following years, the size and velocities of aircrafts increased considerably. The purely muscular actuation of control surfaces therefore became obsolete because incompatible with the level and duration of effort required of the pilot to operate control surfaces. At first, reversible hydraulic actuators were introduced to carry out assistance functions. They were functionally similar to those used in hydraulic steering assistance for automobiles. However, the limits of reversibility were reached very quickly in the face of flutter issues. This is why hydraulic actuators of primary flight controls became non-reversible. In France, such irreversible servo-controls were developed by Jacottet-Leduc in 1950 who built the first three-axis servo-controlled plane: the Sud-Ouest S0 1021 Espadon. The road was now paved for French supersonic fighter jets. The Jacottet-Leduc irreversible servo-control (see Figure 1.9 and left Figure 1.10) servos the position of the cylinder rod to the position of the control rod. For reliability reasons, it is powered by a double hydraulic power source (main and backup) with constant pressure. In case of loss of power, it switches automatically to direct drive mode, backlash-free, between the control rod and the cylinder rod. The servo-control, which consists essentially of bodies of revolution, is mounted in line with the control linkage of control surfaces. The cylinder is of the differential type and its piston constitutes the body of the hydraulic valve. Power metering between the source and the cylinder is carried out by two hydraulic restrictions whose flow section varies as a function of the lift of balls B_1 and B_2. This ball lifting depends on the cylinder rod position relative to the control rod, which performs the subtracting function of the position servo-loop. Hence, the servo-control performs an assisting function by amplifying the effort provided by the pilot to steer the control surface. The same principle of actuator control by the rod to achieve an "in line" assistance function was applied to the servo-controls of helicopter Alouette III, as shown in Figure 1.10. In this implementation, power metering is carried out by a sharp-edged two-way valve rather than a balls valve.

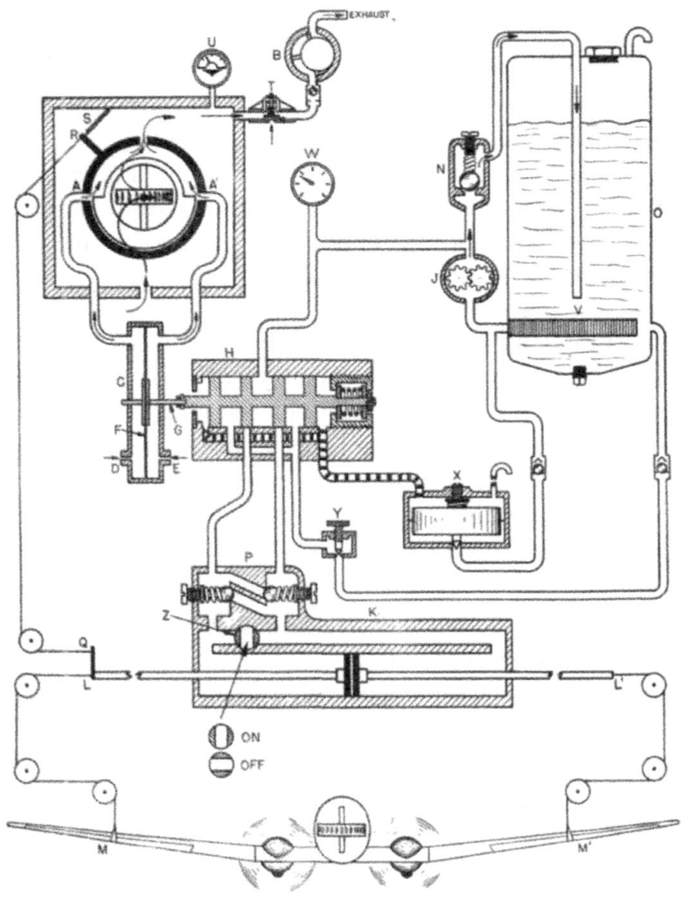

Figure 1.8. *Operating principle of the autopilot in roll [DON 40]*

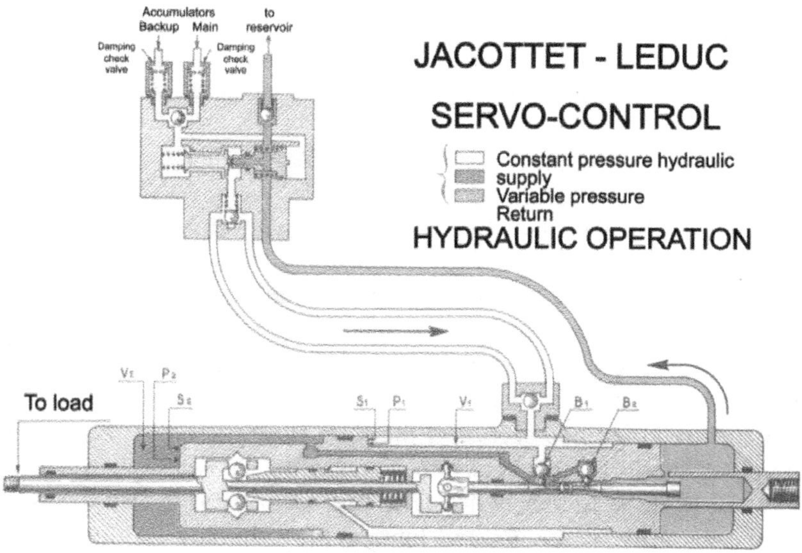

Figure 1.9. *Jacottet-Leduc irreversible servo-control [JAC 54]*

Figure 1.10. *Jacottet-Leduc and Samm hydraulic assistance cylinders (Alouette III)*

The universalization of the use of hydraulics for actuation functions led to a gradual increase of the hydraulic power installed on board: under regular activity mode, a few kilowatts on board the DC3 (1936, 11 tons), 23 kW for the Caravelle (43 tons, 1958), 125 kW on Airbus A320 (1988, 78 tons), 240 kW on Airbus A340 (275 tons, 1993) and 368 kW for Airbus A380 (575 tons, 2007).

This increase in size effect is also well illustrated concerning power requirements for rudder actuation. Stall force and no-load speed are respectively of 3 × 45 kN and 110 mm/s on an Airbus A320, 3 × 94 kN and 135 mm/s on an Airbus A340, 3 × 155 kN and 160 mm/s on an Airbus A340-600 and finally 4 × 225 kN and 100 mm/s on an Airbus A380.

Increase in working pressure

In response to the increase in hydraulic power, while maintaining acceptable mass power and overall dimensions, working pressure has been steadily rising ever since the first use in aeronautics in the 1930s. Indeed, while looking to generate a given force and a given velocity, a raise in pressure enables the displacement of pumps and the section of cylinders to be reduced (see equations [4.1] and [4.2]). And as a consequence, pressure increase also reduces pipe section if the fluid flow velocity is kept constant. On the other hand, it is necessary to increase the thickness of walls, which are required to resist higher hydrostatic stresses. And it is also necessary to further the maturity of sealing and lubrication technologies. In the end, the mass gain mostly shows, in terms of the return lines (and the fluid they contain), that they still operate at the same pressure whereas their diameter can be reduced thanks to the decrease of flow rates that need transmitting.

As shown in Table 1.7, pressure has increased tenfold in a little more than half a century, at the request of spatial and military applications. In the 1980s, serious efforts have been made overseas to assess the benefit of rising the working pressure of military aircrafts to 8,000 psi (552 bar) [JEF 98] [SAE 04]. Knowing the reduction of hydraulic fluid flammability goes hand in hand with this pressure increase, special hydraulic fluids, of type CTFE (chlorotrifluoroethylene), were specially developed [VAN 90]. Compared to the mineral-based fluid Mil-H-83282 commonly used, the combustion heat of fluid CTFE A02 was 7.4 times lower and its auto-ignition temperature was almost double that of the mineral fluid. Unfortunately, its density was increased by 120%, which heavily altered the mass balance. Furthermore, the Bulk modulus was decreased by 11%, which degraded actuator dynamic performance. In the end, the working pressure of fighter jets was frozen at 5,000 psi (350 bar) in the 1990s.

In the early 1960s, the standard working pressure in commercial aeronautics had also been frozen for more than 40 years at 207 bar (3,000 psi). Experience gained in the military and spatial domains made it possible to propagate the pressure increase to commercial transport, albeit with a delay of several years. This delay was in part due to more stringent demands in terms of lifespan, reliability and availability[5]. Since 2007, Airbus with the A380, set a new standard at 350 bar (5,000 psi), which then quickly spread to newer large passenger jets (Boeing 787 in 2011, then Airbus A350 in 2015).

Commercial airplanes	1935 Douglas DC3 35 bar	1959 Caravelle 165 bar	1964 Boeing 727 207 bar	1975 Concorde 280 bar	2007 Airbus A380 350 bar
Military airplanes		1950 Boeing B50 100 bar	1952 Boeing B52 207 bar	1968 Jaguar 280 bar	1991 Dassault Rafale 350 bar
Helicopters	1952 Sikorsky S55 55 bar	1977 Aerospace Super Puma 175 bar	1995 NH industry NH90 207 bar		
Space launchers			Titan 207 bar	1967 Saturn V 250 bar	1996 Ariane V 350 bar

Table 1.7. *Evolution of the working pressure of hydraulic systems in aeronautics*

As a guideline, the weight of hydraulic systems of commercial aircraft typically represents 0.5 to 1.8% [ROS 00] of their maximum take-off weight (MTOW). Regarding the double-deck Airbus A380, it represents 5 tons out of 550 tons and it is estimated[6] that the transition from a working pressure of 207 bar to one of 350 bar made it possible to save between 500 and 1,000 kg out of the overall mass of the hydraulic system.

1.3.4. *Potential advantages and disadvantages of hydraulic technology*

The benefits of transmitting power in hydrostatic form are numerous.

– *High power density of equipment*

For a given piece of equipment (pump or actuator including their power dosing element), the power density can reach 10 kW/kg, in the best case

5 Military airplanes use a hydrocarbon-based fluid whereas commercial airplanes use a synthetic fluid which is more resistant to fire but has restricting properties.

6 The balance of mass consecutive to the increase in pressure is not easy to do because it has many consequences on other systems.

scenario. This is still very high compared to electrical or pneumatic technologies. This advantage is soon dissipated for powers lower than a few kilowatts.

– *Easy generation of large forces at low speed*

Hydraulic cylinders make it possible to generate large forces at low speed in an extremely simple manner. This process involves no other mechanical reducer than a lever arm effect provided directly by the driven load: this is commonly called a direct drive design.

– *High dynamic response characteristics and high acceleration abilities*

In the absence of a high reduction gear ratio, inertia (mass or moment of inertia) of the moving parts of the actuator reflected at the load is in general very weak with regard to inertia of the load to be moved. As a consequence, during transient phases, hydrostatic forces are predominantly transmitted to the load and are not spent to accelerate moving parts of the actuator itself.

– *Easy dissipation of heat generated by energy losses*

Hydraulic fluid is a very good heat transfer fluid. At the component output, it "naturally" extracts heat generated by energy losses caused within the component. In that regard, power transmission via material transport (hydraulic fluid) is an advantage.

– *Easy implementation of secondary functions*

Hydromechanical power transformers, such as cylinders, show excellent mechanical efficiency and weak mechanical inertia. Pressure in the actuator is therefore an accurate representative of transmitted force, even in transient phases. Therefore, secondary functions such as de-clutching, protecting against excess force or even damping at end-stop, can be performed in the hydraulic domain with low weight, small overall dimensions and high reliability.

– *Absence of sensitivity to (or generation of) electromagnetic interferences*

Electromagnetic interferences have no impact on the behavior of hydromechanics. Hydromechanical components do not generate electromagnetic interferences.

– *High technological maturity*

Technological solutions implemented to transmit and dose power have been established for several decades[7] already. Based on the experience feedback collected for years, databases, standards and recommendations have made it possible to master risks over the entire lifecycle of components and systems.

As always, benefits are counterbalanced by a number of drawbacks, some of which are less and less tolerated:

– *Low power density of the distribution*

Supply and return hydraulic lines, as well as the fluid they contain must be able to resist pressure, external impacts and vibrations. Their weight significantly impacts the global power density of hydraulic systems, more strongly so in large aircrafts that may comprise several kilometers of hydraulic lines. Two examples help illustrate this drawback. The first concerns Airbus A380. For this type of very large aircraft, it appears that the weight of hydraulic equipment typically constitutes 25% of the total weight of the hydraulic system. Pipes, clamping and hydraulic fluid account for the remaining weight. The second example deals with Boeing B777. The main hydraulic pump of the Boeing delivers 182 l/min at a working pressure of 207 bar and has a dry mass of 18.2 kg. In order to transmit this power, several meters of pipes are required. One meter of supply and return pipes typically weight 1.5 kg including the fluid they contain (low pressure pipes made of aluminum and high pressure made of titanium, average fluid speed 7 m/s). Hence, 12 m of lines, sized to allow the transmission of all the power supplied by the pump, weight, with their fluid, the same as the (dry) pump itself!

– *Fluid conditioning requirements*

The hydraulic fluid must be conditioned. Additionally, any assembly or maintenance task imposes harsh constraints regarding security, pressurization, pollution, bleeding and leakage collecting.

7 However, this does not mean that there is no room for improvement!

– *Difficult power management and low re-configurability*

The cost of reducing risks of propagation of fluid pollution and leakage translates into severe requirements that impose independent and segregated redundant hydraulic systems. This considerably hinders power management and limits reconfiguration possibilities for power networks.

– *Integration restrictions*

Bending radii of hydraulic lines are limited by the concentration of mechanical stresses they generate. Moreover, the tolerance for accidental leakage forces the minimization of fire hazard in the vicinity of hot parts (engines) as well as spraying hazard in the cabin. These restrictions have to be added to the previously discussed independence and segregation requirements. In the end, those constraints heavily impact the routing of hydraulic networks and the location of equipment.

– *Aggressiveness of the hydraulic liquid*

Hydraulic liquids, in particular those used in commercial aviation, are skin and eye irritants. They attack numerous different materials and pollute the environment. This aggressiveness is a strong argument to the detriment of hydraulics.

– *Power metering by energy dissipation*

Metering of the amount of power transmitted to driven loads is, in the vast majority, carried out by power dissipation through hydraulic valves. Power is therefore consumed as if generating full force, regardless of the effort requested by the load.

– *Permanent internal leakage*

In order to ensure the smooth and accurate operation of hydraulic equipment, it is necessary to tolerate permanent internal leakage, for example, to implement dynamic sealing or at servovalve pilot stages.

Finally, in comparison to other technologies, the benefits still very clearly outweigh the disadvantages of hydraulic technology. However, the lightning-speed development of electrical technologies (power electronics, electric machines, controls) has already led to successful flight-worthy developments, although only with the purpose of partially performing critical actuation functions (for instance for the actuation of 4 out of the 14 spoilers on Boeing B787). It is difficult to accurately compare technologies because it requires assessing the gain for the final user, for example, the passenger and the airline company. To carry out this comparison process fairly, it is important to remember not to conceal intrinsic benefits of "conventional" solutions and to research in depth drawbacks specific to "newer" solutions. Goals and restrictions should also be considered from a global, lifecycle point of view, regarding new services that can be provided as well as risk, costs and environmental impact minimization.

1.3.5. *Overall hydraulic circuit architecture*

Figure 1.11 shows a hypothetical power architecture of an aerospace hydraulic system. Without going into details, the first step is to visualize the architectural concepts applied to carry out the most generic functions of hydraulic power transmission, which are:

– convert power between the hydraulic and mechanical domains;

– transport hydrostatic power using properly conditioned hydraulic fluid;

– dose the power transmitted to the load;

– manage (distribute/transmit) power as a function of life stages, reliability constraints and response to failure.

These four major functions will each be addressed in one of the following chapters of this volume. A final chapter is dedicated to the integration of components, equipment and hydraulic systems.

As shown on Figure 1.11, hydraulic schematics that will be introduced in this book will respect as much as possible standard symbols SAE-AS1290-B [SAE 11]. Additionally, colors will be used to differentiate high pressure lines (red), low pressure lines (dark blue), suction (light blue) and actuators (green).

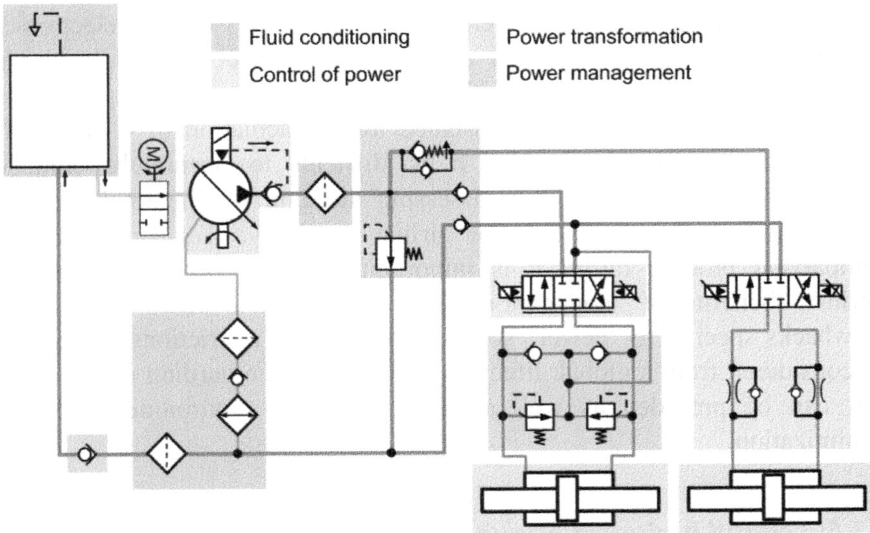

Figure 1.11. *Typical hydraulic system architecture. For a color version of the figure, see www.iste.co.uk/mare/aerospace1.zip*

2

Reliability

This chapter briefly outlines the general concepts of reliability with a particular focus on redundancy architectures. It is important to pay close attention to the specific technical vocabulary used in this chapter.

2.1. Risks and risk acceptance

Any system can potentially be *dangerous*, meaning it can cause harm to people, goods or the environment. Risk management consists of controlling the probability of systems being harmed when exposed to danger. For this purpose, systems have to be given minimal safety properties depending on their hierarchy rankings in order to keep risks under control. And these properties have to be maintained throughout their lifecycle. Risk acceptance depends on its *criticality*. The criticality of a risk can be defined as a risk that takes into account the extent of damages caused (their *gravity* or *severity*) and their likelihood (probability of occurrence). Optionally, it can sometimes include risk detectability. It is understandably accepted that an event with minor consequences occurs more frequently than an event with catastrophic consequences. A more algebraic definition of criticality follows:

$$\text{Criticality} = \text{Gravity} \times \text{Occurrence} \times \text{Detectability}$$

Risk management is most often based on acceptability: this concept can for example be defined by Table 2.1, where the higher level of criticality is considered catastrophic.

Risk gravity	Negligible	Marginal	Critical	Catastrophic
Risk consequence / Risk probability	Discomfort, negligible damages to the system or the environment	Injuries and minor inability to work, minor impact on the system and the environment	Serious injuries and inability to work, heavy impact on the system and the environment	Death and severe damages to the system and the environment
Frequent (very often during the product's life)	Tolerable	Unacceptable	Unacceptable	Unacceptable
Likely (often during the product's life)	Acceptable	Tolerable	Unacceptable	Unacceptable
Rare (fairly often during the product's life)	Acceptable	Acceptable	Tolerable	Unacceptable
Extremely rare (a few times during the product's life)	Acceptable	Acceptable	Acceptable	Tolerable
Extremely unlikely (probably never during the product's life)	Acceptable	Acceptable	Acceptable	Acceptable

Table 2.1. *Risk matrix: risk acceptability as a function of its gravity and its frequency [FAA 14]. For a color version of the table, see www.iste.co.uk/mare/aerospace1.zip*

Concerning the design of civil aircraft and their systems, various standards (such as standard ARP4754A/ED-79A) lay down the rules that must be applied in order to design reliable systems and aircraft. Generally speaking, the *reliability* of a function is defined as the ability to perform this function during a given amount of time, in a given environment and for a given purpose.

However, it is extremely difficult to formally prove reliability regarding design processes, in particular for information systems. Therefore, emphasis is primarily placed on this design process. For example, standards D0-254/ED 80 for equipment and DO-178B/ED-12C for software introduce five qualitative Design Assurance Levels (DAL) applied to functions

(F-DAL for Functional DAL) or to their physical constituents (I-DAL for Item DAL).

For energy systems, it is much easier to prove, by means of a calculation, that architectures and sizing make it possible to satisfy reliability requirements. In this situation, the *acceptability threshold* must then be defined numerically for each degree of gravity, at the maximum permissible occurrence probability. In order to achieve this, failure rate h is introduced. It is defined as the probability of failure expressed with respect to provided service quantity u. In the field of aeronautics, the failure rate is therefore expressed per flight hour (/FH)[1].

In practice, the challenge lies in quantifying acceptability. This is particularly true when defining the maximum tolerated failure rate for a catastrophic event, in order to make this event extremely unlikely. Concerning commercial aeronautics, this threshold is tied to in-service experience feedback. Typically, it is considered that, at most, one catastrophic event occurs every million flight hours[2] and that only 10% of disastrous accidents were caused by the airplane itself. Furthermore, on a commercial aircraft, it is estimated that 100 different and independent technical failure causes have catastrophic consequences, each having approximately the same probability of occurrence. In the end, these considerations have led to setting $h = 10^{-9}$/FH as the maximum tolerated failure rate for each technical failure of the airplane with catastrophic consequences. The maximum acceptable failure rate is then increased 100 fold to define the successive degrees of gravity, as shown in Table 2.2. Therefore, for a given component (device, equipment, sub-system or system), the tolerated quantitative probability of failure depends on its potential to contribute to the catastrophic risk at the airplane level.

1 Provided services can be expressed in other units, such as in million revolutions for ball bearings.

2 Over the period 2004–2014, the average accident rate for cutting-edge occidental commercial aircrafts (electrical flight controls with flight envelope protection) is as follows: 0.23 airplane were destroyed and 0.11 human lives were lost per million flights [AIR 14]. When taking into account all commercial jets, of all generations, these values rise to 0.77 and 0.33, respectively [BOE 15].

Quantitative probability	10^{-3}/FH	10^{-5}/FH	10^{-7}/FH	10^{-9}/FH	
Qualitative probability	Frequent	Probable	Remote	Extremely remote	Extremely improbable
Gravity	Minor		Major	Hazardous	Catastrophic
Consequences on the airplane and its passengers	Nuisance	Operational limitations Emergency procedures	Significantly reduced safety margins Crew experiences difficulties handling the situation Injury to passengers	Heavily reduced safety margins Increased workload for the crew Serious or deadly injuries to a number of passengers	Many dead Loss of aircraft

Table 2.2. *Acceptability thresholds and consequences on passengers [FAA 14]*

It is difficult to form a mental representation of failure rates. In order to get a better idea of what h represents, it can be helpful, for example, to notice that a failure rate of $h = 10^{-9}$ per flight hour is equivalent to tolerating one event every billion flight hours. Considering a fleet of 2,000 Airbus planes of the A320 family, flying 10 hours a day and 300 days a year, then 1 year in service corresponds to six million flight hours. It will therefore take as a long as 167 years to reach one billion flight hours. This consideration gives a better understanding of the "extremely improbable" label.

In addition to the threshold values defined above, a generic requirement applies in all cases for commercial airplanes: *no single failure should lead to a catastrophic event.*

2.2. Response to failure

The architecture and the technological choices selected to design a system depend heavily on the criticality of the functions the system performs. Where appropriate, the system can be designed to avoid failure (resistance to failure) or to control failure in case one of its components has a fault (tolerance to failure).

2.2.1. *Resistance to failure*

In a number of very specific cases, it is possible to prove that the product will never fail during its operational life (safe-life). This phenomenon can be referred to as resistance to failure. This property is obtained by over-sizing and/or by periodically replacing components. This approach is essentially used in structural parts such as landing gear struts, for which safety margins are applied regarding resistance and mechanical fatigue. The following are, for example, imposed:

– no permanent deformation, after applying the limit (or proof) load met in normal service conditions;

– no rupture for at least 3 s when subjected to the ultimate load. The ultimate load is defined by applying a safety coefficient to the maximum in-service load (for example 1.5 for a 50% margin). For a loading corresponding to the ultimate load, residual or permanent deformation after removing the load is tolerated.

It is acceptable for the proof of resistance to failure to be provided by combining the use of design software certified for numerical calculations, actual tests and regular inspections.

	$\frac{\text{Proof pressure}}{\text{service pressure}}$	$\frac{\text{Ultimate pressure}}{\text{service pressure}}$
Pipes and fittings	1.5	3
Hydraulic hoses	2	4
Components with pressurized gas		
High pressure (accumulators)	3	4
Low pressure (reservoirs)	1.5	3
Other elements	1.5	2

Table 2.3. *Examples of safety margins for hydraulic components [FAA 01]*

2.2.2. *Tolerance to failure*

Concerning most systems of the aircraft, over-sizing is not advisable, not possible or not enough to reach the degree of reliability required. This is particularly true for safety-critical systems of commercial aircrafts since no single fault should lead to a catastrophic event. Therefore, systems must be given tolerance to failure properties in case one or more of their elements[3] fails. For a system, different generic failure types resulting from the fault of one of its elements can be identified:

– Hardover. The quantity to be outputted from the system takes an extreme value (for example, the cylinder rod comes out and reaches its end-stop). For flight controls for example, hardover failure is feared because the aerodynamic forces on the rudder become excessive and quickly lead to rupture.

– Fail-open. The system output is opened (for example, the ball-joint linking the cylinder rod to the load is opened and it is no longer possible to transmit force). For flight controls for example, fail-open is feared. Depending on the control surface, it can lead to flutter or to loss of longitudinal balance in case the actuator of the trimmable horizontal stabilizer fail-opens. Conversely, concerning nose landing gear steering, opening allows the wheel to self-align. Steerability on the ground is then carried out by differential left/right braking.

– Jamming (or seizure). The system output is frozen (for example, the cylinder rod is jammed in the cylinder body and it is no longer possible to transmit motion). For certain flight controls such as the trimmable horizontal stabilizer, jamming is the preferred failure mode. This is because longitudinal balance is hardly affected and can be corrected by the elevators.

– Erratic operation. The quantity to be outputted from the system takes erratic values. It then becomes very difficult to compensate for this situation, which can be very dynamic.

For a system that tolerates the presence of a fault in one of its elements, it is possible to specify its response type depending on the function it performs

3 From here on in, the word "element" will be used to mean components as well as information, power or matter flows.

and the nature of the fault involved (the feared event in question). Figure 2.1 introduces generic response types that enable the system to:

– remain active, meaning it continues to perform its function, possibly with reduced performance (fail-active);

– or stop functioning all together without disturbing other systems or their environment (fail-safe).

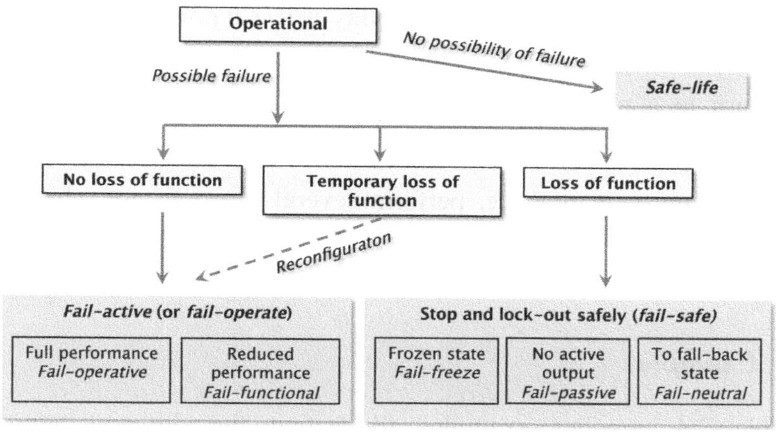

Figure 2.1. *Response of a system to the failure of one of its elements*

2.2.2.1. *Fail-active system*

When loss of function is not tolerated, system performance must be ensured, either completely (in which case the system is said to be fail-operative) or in a reduced manner (in which case the system is said to be fail-functional). This performance requirement can also apply in the event of a double failure, in which case the system is said to be double fail-operative or double fail-functional. If it is possible to sustain a loss of function for a short while, the failure triggers a system reconfiguration. Following this, the system can once again perform its function with full performance or at least with reduced performance.

2.2.2.2. *Fail-safe system*

If loss of function is tolerated, the system must stop functioning without disturbing other systems or their environment. Such a system is said to be fail-safe. Depending on the situation, the system must then:

– remain frozen in the configuration it was in when the failure occurred, in which case it is called a fail-freeze system;

– let itself be dragged along by other systems, optionally while dissipating energy to contribute to stability if it is a power system, in which case it is called a fail-passive system; and

– return to a predetermined neutral configuration, in which case it is called a fail-neutral system.

Systems must, in practice, perform several functions for which feared events and expected responses can differ. As a consequence, various types of response to failure are often combined within a single system.

2.2.3. *Examples*

Table 2.4 provides examples illustrating the different approaches introduced previously. These examples are, for the most part, focused on actuation functions. It is worth noticing that, for the same equipment, response to failure will be different depending on the feared failure event involved (see elevator control example) and on the number of faults encountered (see aileron controller example).

2.3. Redundancy

Redundancy is used in order to provide a system with tolerance to failure properties. It consists of duplicating system elements or adding backup (or standby) elements. Redundancy therefore makes it possible to maintain the system in an active state, meaning it remains able to perform normally in the presence of one or more element faults. Elements, which together form an independent unit, are defined as a channel. A system with a single channel, in other words non-redundant, is called a *simplex*.

System	Examined function	Example of a feared event	Response to failure
Gear strut	Transmit force between the wheel and the airframe	Mechanical rupture	Safe-life No fault tolerated
Helicopter swashplate control	Impose the collective and cyclical pitch of the main rotor blades	Failure of one of the three actuators	Fail-active/fail-functional No transient loss of function, reduced performance tolerated
Aileron control (two actuators per aileron, only one active at a time)	Steer the aileron according to the pilot's setpoint	Failure of an element of the normal actuator (controller, pressure supply, servovalve, sensor, etc.)	Fail-active/fail-operative Transient loss of function. Then performance returns to its initial value after activation of the backup actuator
		Then, failure of the backup actuator, which switched to active mode	Fail-safe/fail-passive Permanent loss of function, the two actuators work as dampers
Airbus A320 elevator control	Steer the elevator according to the pilot's setpoint	Loss of hydraulic power	Fail-safe/fail passive The actuator switches to damped passive mode
		Loss of electrical control signals	Fail-safe/fail-neutral The actuator returns to a neutral position (no elevator steering)
Trimmable horizontal stabilizer (THS) pitch control	Ensure equilibrium of forces applied to the airplane about the pitch axis (pitch trim)	Failure of the THS actuator	Fail-safe/fail-freeze The actuator becomes stationary and freezes the THS

Table 2.4. *Examples of response to failure*

In terms of actuation, redundancy of the actuation function is implemented by summing either forces or displacements outputted by each channel. The summation of displacements is hardly ever used. Figure 2.2 illustrates the concept of force summation applied to actuators with multiple hydraulic power channels. There are two two-channel solutions. First, the *tandem* solution associates two pistons on the same rod. This is equivalent to having two rods mounted in series (geometrically). Second, in the *duplex* solution, the two cylinder rods are connected at the stage of transmission to the load. This is equivalent to having two rods mounted in parallel (geometrically). The main difference between these two solutions is their overall dimensions: the first is rather thin and long, while the other is shorter but wider. *Triplex* and *quadruplex* solutions sum efforts made by three or four rods at the stage of transmission to the load; the actuator body is made out of the assembly of three or four distinct bodies.

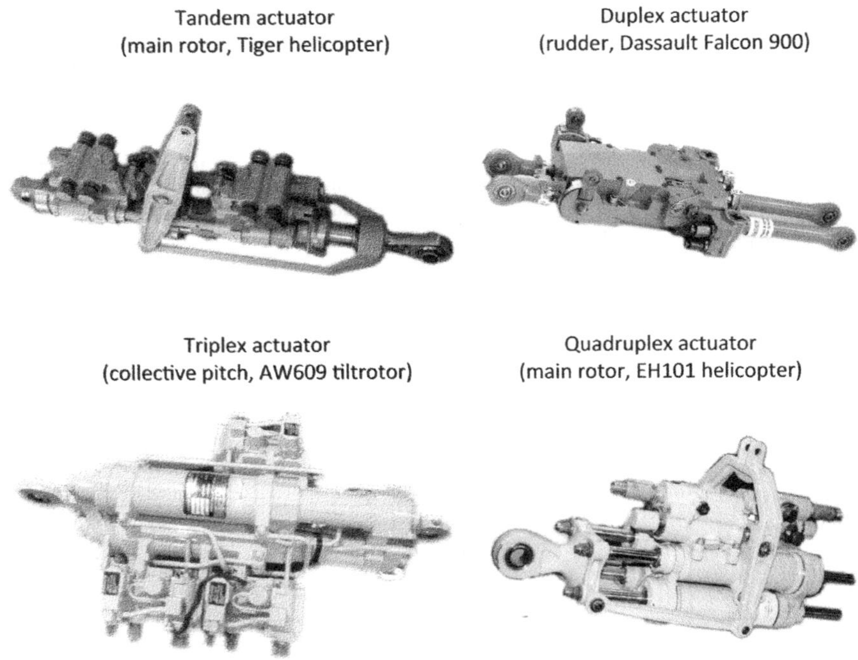

Figure 2.2. *Force summation in flight control actuators*

Redundancy types are difficult to classify due to the wide range of implemented solutions and due to differences in technical language used in the various areas of engineering. References [TAY 76] and [TAY 80] deal with redundant architectures for hydraulic servo actuators. They have constituted a useful source of information to help structure the following sections.

Figure 2.3 suggests a possible classification for the main types of redundancy used in actuation systems. It highlights the two major redundancy families – static and dynamic – and their combination, which is described as hybrid redundancy. An example of such hybrid redundancy is given by the rudder actuation of Embraer ERJ 145 [BEA 08], which is carried out by the summation of forces outputted by two hydraulic cylinders. As long as the relative speed of the aircraft is lower than 135 kt (250 km/h), the two cylinders are active simultaneously (active/active mode). Above this speed, only one cylinder remains active (active/passive mode).

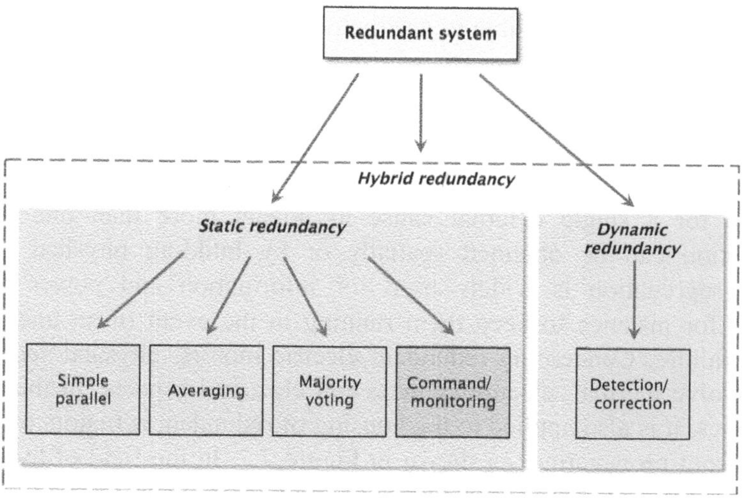

Figure 2.3. *The main types of redundancy*

There are other redundancy architectures that are specific to signal processing systems. Additionally, preferred technical language tends to differ between the domains of power and information.

Redundancy architectures introduced in Figure 2.3 are detailed in the following sections. For the sake of simplicity, it is standard practice to consider only one common input and to assume it is distributed out evenly to every channel. In practice, inputs often have very different origins even though they convey similar quantities (for example, commands developed by separate computers).

In order to take full advantage of redundancy, systems are given specific properties: *independence*, *diversity* and *segregation*.

– Independence ensures that the failure of one channel does not spread to other channels.

– Diversity prevents the same internal fault from appearing in all channels. This solution relies on giving different design features to every channel, especially computer systems. Therefore, the aim is to pick different devices (such as computers), different programming languages, different wiring and even different engineering teams from different geographical locations. Regarding power transmission systems, diversity is implemented by involving different physical principles. For instance, in order to drive a hydraulic pump, it is possible to use the power of either the main engine, an electric motor, a wind turbine or even an air turbine.

– Segregation consists of the absence of shared elements between channels. This property therefore removes common mode failures. Additionally, segregation contributes to independence. Therefore, it is not possible for a single external cause to impact more than one channel. Segregation can be obtained spatially or by building physical barriers. Spatial segregation is widely used for information and power network routing, for instance to keep them running in the event of an uncontained engine failure. Concerning redundant electric motors, physical segregation may involve thermal, electromagnetic and electrical aspects. In the field of mechanics, it is also applied to the housing of redundant actuators to prevent cracks from propagating, as shown in Figure 2.2. In the field of hydraulics, segregation prevents the propagation of fluid pollution and leakage.

2.3.1. *Static redundancy*

In systems with static (or active or parallel) redundancy, channels are duplicated and all are simultaneously active under normal circumstances. The function is performed by summing the outputs of each of the channels. In the event that one of its elements has failed, the concerned channel may or may not be isolated (or masked). Redundancy is described as *purely parallel* when the failure rate of an active channel does not depend on the status – functional or failed – of the other channels. This situation is mostly only met in information systems, where there is no direct interaction between the different channels mounted in parallel. Conversely, redundancy is described as *shared parallel* when the failure rate of an active channel depends on its contribution to the function, which increases when other channels are failing. This behavior is typical of power transmission systems, where failure rate is tied to the amount of power generated. Furthermore, in power systems, the various channels mounted in parallel must deliver identical outputs in order to avoid force-fighting. It is therefore crucial to ensure they are synchronized. This can, for example, be achieved by adding functions or adjustment procedures to equalize outputs of simultaneously active channels. This thereby avoids loading channels unnecessarily, which would negatively impact their reliability. Regarding actuation, a typical example of static redundancy is the actuation of a control surface by two simultaneously active actuators. In practice, the two actuators differ as a consequence of their manufacturing process (sourcing, manufacturing and adjustment tolerance margins) and their integration (differences in temperature, structural compliance, force distribution on the load). As a result of these differences, each actuator with a position servo-loop is actively trying to impose a specific position to the load. Each actuator is therefore fighting against the other actuator, which is also trying to impose its position. In the end, the position reached is satisfactory but the actuators, the structure and, depending on the circumstances, the power networks are unnecessarily solicited: this fight for control is commonly called *force-fighting*.

The main benefits and drawbacks of static redundancy are summarized in Table 2.5.

Static redundancy	
Advantages	Disadvantages
No loss of function, even during the transient phase following the occurrence of a failure.	
No need to have functions of detection and isolation of the failing element given that it is corrected "naturally".	In principle, monitoring signals are absent.
For power systems, elements contribute fairly (in theory) to generating the controllable variable (e.g. position). Wear of the elements is identical. Power usage is shared between the networks associated with each element[4].	Precautionary synchronization measure to avoid or limit force-fighting between different channels. This impacts sizing, controls and maintenance.
No non-redundant elements.	

Table 2.5. *The main advantages and disadvantages of static redundancy*

4 This remark is poorly suited for hydraulic actuators when power metering is achieved by variable hydraulic restrictions and when power is supplied at constant pressure: by the conceptual principle, restrictions must waste all supply pressure in excess with respect to load demand. As a consequence, and unless stated otherwise, when such actuators are associated in static redundancy, total power consumption is twice that of a simplex solution.

As for power systems, there are cases where elements are associated in parallel, not for redundancy purposes, but to reduce the power each element has to supply. This makes it possible to meet integration constraints (for example, geometrical envelope or weight to manipulate for maintenance) or simply technological availability constraints (no single element is capable of supplying the whole amount of power required).

2.3.1.1. *Simple parallel redundancy*

A system with simple parallel redundancy (or active/active) is built by associating two channels in parallel and by summing their outputs (for instance flow rate summation, force summation or displacement summation), as shown in Figure 2.4. It is imperative that each of the channels has a response to failure of the fail-safe type, suited to the application. Each channel must also be sized depending on the expected response of the system. If the system has to be of the fail-operative type, a channel must be able to deliver full performance all by itself in case the other channel fails. As for the simple parallel architecture, the function is performed from channel 1 and/or 2. This may be seen as equivalent to the two channels being connected via an "inclusive OR" gate.

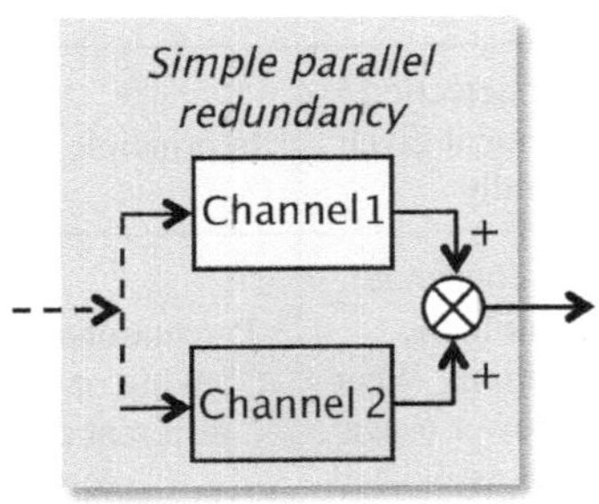

Figure 2.4. *Simple parallel redundancy*

2.3.1.2. *Redundancy by averaging*

In order for the system to remain operational when one of its elements is actively failing, at least three channels must be mounted in parallel and their authority must be limited. This concept, described in Figure 2.5, is called *averaging*. In the event that one channel fails, system performance is reduced (by one-third for three channels). In the event of an active failure, a bias appears on the output. Consequently, the response of the system with redundancy by averaging to an active fault of one of its channels is inevitably of the fail-functional type. It is worth noting that a limitation is established to allow remaining channels to carry out the function despite one channel failing.

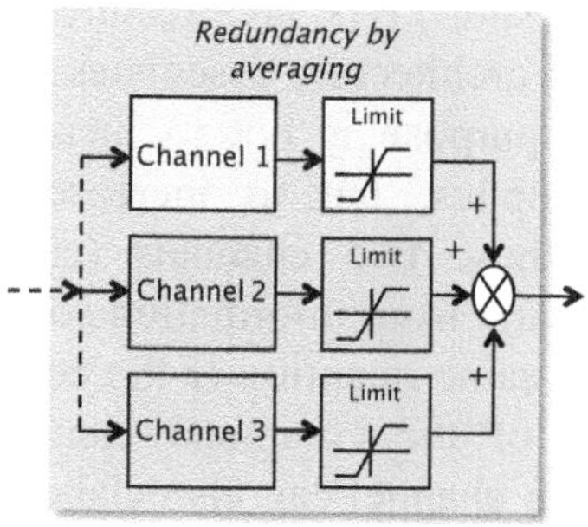

Figure 2.5. *Redundancy by averaging*

2.3.1.3. *Majority voting redundancy*

The majority voting concept (see Figure 2.6) consists of feeding back the output to each of the inputs, in order to reduce the bias caused by the failing element in averaging power transmission systems. This bias can even be entirely cancelled by taking advantage of the presence of an integrating effect downstream the summation of channel outputs (for instance the position of a cylinder rod is, broadly speaking, the integral of the flow rate supplied to cylinder chambers). When it is used, this solution is often combined with a detection–correction concept.

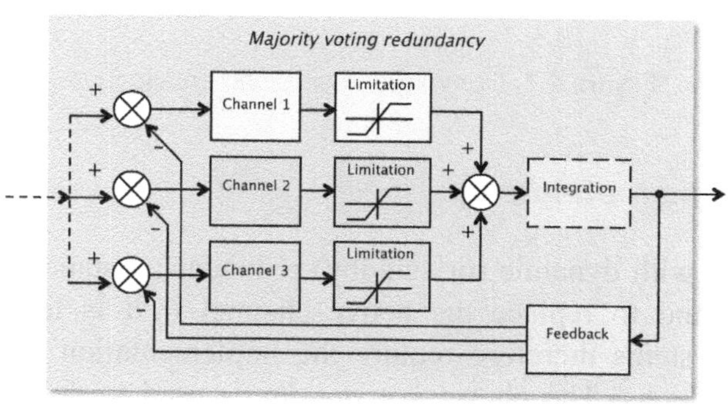

Figure 2.6. *Majority voting*

2.3.1.4. *Command/monitoring*

Concerning signal processing, a widely used approach for flight controls consists of implementing two computers according to a command/monitoring (com/mon for short) type architecture [TRA 06], as described in Figure 2.7. Although this architecture associates two simultaneously active channels in parallel, its purpose is not to increase the reliability of the function to which it applies, but to increase fault detectability. The computers of each of these two channels (command and monitoring) elaborate a control command from information sources that are different but still representative of the same quantity. If the control signals are identical, then the control signal of the command channel is transmitted to its recipients. The monitoring channel can also elaborate commands validating the active mode of controlled elements. This control signal is therefore only used for validating. For this architecture, the function is only performed on the condition that both channels function properly. This thereby sets up a "logical AND" condition on the operation of the two channels.

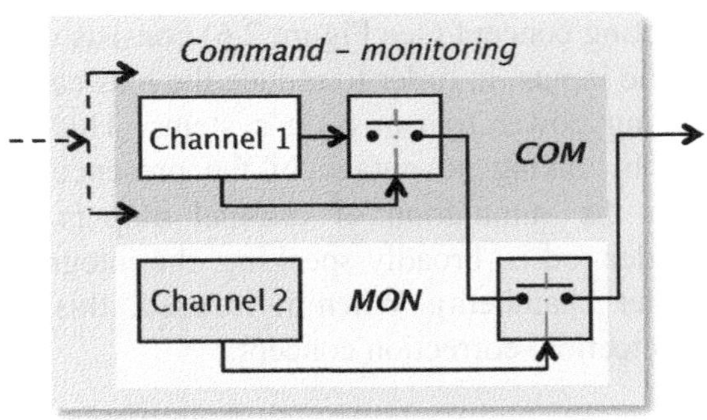

Figure 2.7. *Command–monitoring architecture*

2.3.2. *Dynamic redundancy*

In systems with dynamic (or standby) redundancy, standby channels are used on demand to replace the active channel if it is failing. Dynamic redundancy systems therefore require the implementation of one or more switching functions. The phrase *hot standby* is used to describe a standby channel that is powered when not activated. *Cold standby* describes the opposite. Hot redundancy has the benefits of offering quick access to the standby function and preventing the presence of hidden faults. Unfortunately, constantly supplying power comes with high energy consumption and potentially ageing, for example as a consequence of temperature or pressure. And this is the case despite the fact that no power is generated at the output in standby mode. Cold redundancy does not have this disadvantage. However, it delays function take over by the standby channel and it increases the chances of hidden faults. This is because, in this situation, the channel cannot be tested when it is in standby mode. A distinction is made between *warm redundancy* and hot redundancy if the channel is not loaded at the output when it is powered and in standby mode.

The main advantages and disadvantages of dynamic redundancy are summarized in Table 2.6. It is also worth knowing that in certain applications, active channel functions and standby functions are interchanged, for example every other flight, in order to fairly distribute usage between the two channels.

Dynamic redundancy	
Advantages	Disadvantages
	Time taken to detect the fault, take the decision to switch, isolate the normal channel and have the standby channel take over the function causes a transient loss of function.
The detection (diagnosis) function implemented to order switching to the standby channel enables a monitoring signal to be generated.	Diagnosis and switching functions are not generally redundant. This has a negative impact on the global reliability of dynamic redundancy.
For power systems, given that only one channel is active at once, there is no desynchronization issue.	The channel in service must be able to deliver full performance all by itself. The normal channel ages quickly, as opposed to the standby channel.
If the standby channel is sized appropriately, there is no performance loss between the normal mode and the standby mode.	The entire power needed to perform the function is collected from only one power network.
	It is necessary to take precautions against possible hidden faults in the standby channel, which is not used in normal mode (dormant fault).

Table 2.6. *The main advantages and disadvantages of static redundancy*

Detection–correction

The detection–correction (or active/standby) architecture, as described in Figure 2.8, typically includes:

– two distinct channels that are never both active at the same time, each being able to perform the expected function;

– an additional diagnosis function to detect the fault of the active channel. The diagnosis is generally carried out by comparing the generated output with the expected output. The latter is obtained with the use of a model fed with the same control signals as the two channels. The model, which in the past used to be an actual small-scale model of the set-up, is now a numerical simulation model. When the diagnosis detects the failure of the normal channel, a switching command is emitted: upon switching the failing normal channel is isolated and the standby channel is activated. The corrective fix is considered complete once the performance returns to normal levels thanks to the standby channel;

– an additional switching function, which connects the output to the channel selected by the diagnosis function. Switching from the normal channel to the standby channel is called *reversion* or *changeover*.

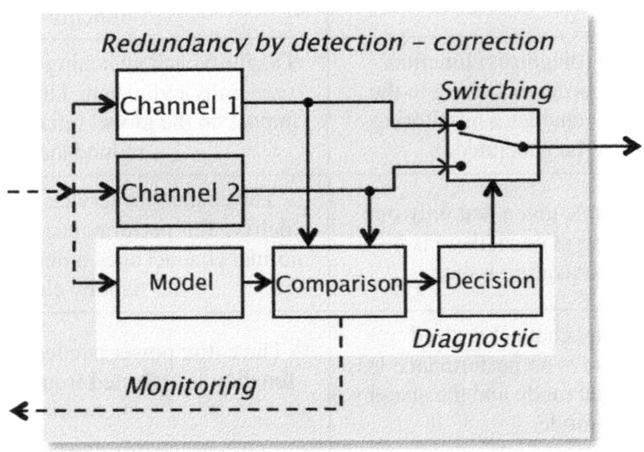

Figure 2.8. *Detection–correction*

In some cases, the performed function contributes to a more global function which requires multiple redundancies involving different concepts (for example, regarding power supply, power metering, measurements, controls, etc.). Then, in these situations, it is preferable to always use the same channel in normal mode. The second channel can then be described as a backup channel. However, certain applications make it possible to alternatively assign the role of active channel to either of the two channels. This allows fatigue and ageing to be shared in the best way.

Concerning detection–correction type redundancy, the function is performed from channel 1 or channel 2, meaning it is carried out by a logical "exclusive OR" gate connecting the two channels.

The concept of detection–correction can also be implemented with more than two channels but it quickly becomes complex.

2.4. Feared events and failure rates in actuation

Table 2.7 gives orders of magnitude for the failure rates of various elements of an actuation system. The various faults within a system all impact the system with more or less severity. This is why failure rates are defined with respect to a given failure mode. For instance, for an actuator, it is possible to consider failure by jamming, by hardover or by opening.

Failure rate		λ/FH
Single-source hydraulic power network		10^{-5}
Hydraulic system		10^{-4}
Two-source electrical network		10^{-8}
Mechanical flight control simplex	opening	10^{-7}
	jamming	$<10^{-7}$
Hydraulic actuator simplex	cylinder jamming	$<10^{-7}$
Hydraulic actuator simplex	hardover	10^{-5}
Hydraulic actuator simplex	global	10^{-6}
Electro-hydraulic actuator		10^{-5}
Electrical actuator simplex		10^{-5}
Computer		$0.5\ 10^{-4}$
Sensor		$0.5\ 10^{-6}$

Table 2.7. *Orders of magnitude of failure rates for various actuation elements and failure modes*

2.5. Fundamentals of reliability calculation

This section provides the main definitions and the bases for the calculation of reliability [OCO 12] that are fundamental to dealing with reliability engineering.

2.5.1. *Variables used in reliability calculation*

In order to define the variables involved in reliability calculation, consider a set of N products in service. The service quantity u is expressed in different units depending on the application: for example, flight hours for an aircraft, millions of revolutions for a bearing, distance travelled for a cylinder rod seal. For each value of the service quantity u_k, the number of elements n_k that have failed over the interval of one unit of service is recorded. The following quantities can then be specified:

– Probability of failure density $f(u_k)$:

$$f(u_k) = n_k / N. \qquad [2.1]$$

– Probability of failure $F(u_k)$ for a service quantity u, which is the integral of the density of failures since the activation with respect to u:

$$F(u_k) = \int_0^{u_k} f(u)du. \qquad [2.2]$$

– Probability of success $R(u_k)$ (commonly known as reliability) for a service quantity u_k:

$$R(u_k) = 1 - F(u_k). \qquad [2.3]$$

– Failure rate (or danger rate) $h(u_k)$, for a service quantity u_k:

$$h(u_k) = f(u_k) / [1 - F(u_k)]. \qquad [2.4]$$

– Mean Time To Failure (MTTF), which quantifies the mean duration of smooth operation without maintenance or repairs since activation:

$$MTTF = \int_0^{\infty} R(t)dt. \qquad [2.5]$$

– Mean Time Between Failure (MTBF), which is expressed as a function of Mean Time To Repair (MTTR):

$$MTBF = MTTF + MTTR. \quad [2.6]$$

Here, the mean time taken to carry out repairs includes all maintenance tasks (disassembly, transport, provisioning, etc.)

– Availability A is defined as the ratio between smooth operation duration and the total expected service time. Inherent availability A_i, meaning when all maintenance means are instantaneously available, is expressed as:

$$A_i = \frac{MTBF}{MTTF + MTTR}. \quad [2.7]$$

The following example illustrates these calculations for a total of $N = 285$ products. Table 2.8 gives hypothetical measurements as well as the results of the above calculations when applied to them. The evolution of variables as a function of service quantity is plotted in Figure 2.9.

Measured		Calculated using measurements				
u service units	Number of n_k failures during 1 unit of service	Number of accumulated failures from the start	Probability density of failure $f(u_k)$	Probability of failure $F(u_k)$	Probability of success $R(u_k)$	Failure rate $h(u_k)$
1	40	40	0.1404	0.1404	0.8596	0.1633
2	25	65	0.0877	0.2281	0.7719	0.1136
3	17	82	0.0596	0.2877	0.7123	0.0837
4	12	94	0.0421	0.3298	0.6702	0.0628
5	10	104	0.0351	0.3649	0.6351	0.0552
6	9	113	0.0316	0.3965	0.6035	0.0523
7	8	121	0.0281	0.4246	0.5754	0.0488
8	8	129	0.0281	0.4526	0.5474	0.0513
9	8	137	0.0281	0.4807	0.5193	0.0541
10	8	145	0.0281	0.5088	0.4912	0.0571
11	8	153	0.0281	0.5368	0.4632	0.0606
12	8	161	0.0281	0.5649	0.4351	0.0645
13	8	169	0.0281	0.5930	0.4070	0.0690
14	8	177	0.0281	0.6211	0.3789	0.0741
15	10	187	0.0351	0.6561	0.3439	0.1020
16	18	205	0.0632	0.7193	0.2807	0.2250
17	30	235	0.1053	0.8246	0.1754	0.6000
18	50	285	0.1754	1.0000	0.0000	–

Table 2.8. *Example of reliability data analysis*

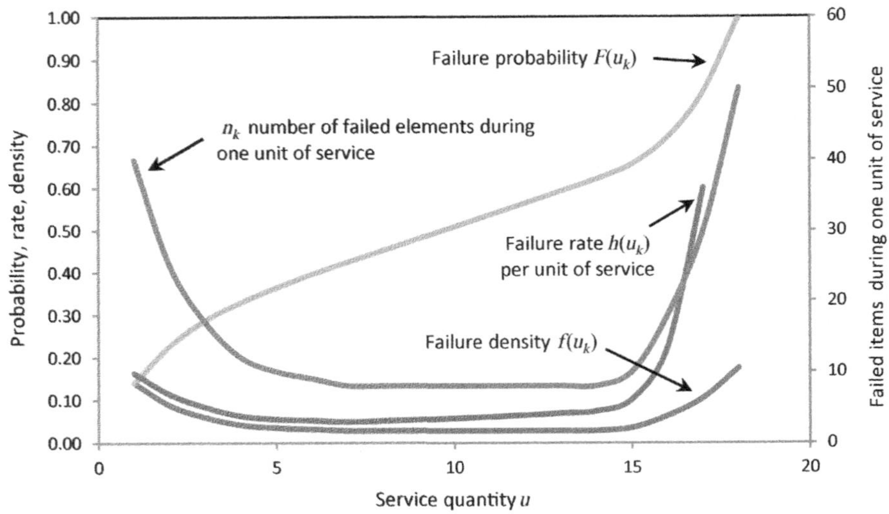

Figure 2.9. *Evolution of reliability variables as a function of service quantity. For a color version of the figure, see www.iste.co.uk/mare/aerospace1.zip*

REMARKS.–

1) If the failure rate is constant and equal to λ during the product's life and if the repair time (MTTR) is negligible compared with the mean time to failure (MTTF), then $MTBF = 1/\lambda$.

2) Experience feedback allows databases to be built in order to statistically predict reliability from variables defined by the equations above. For example, out of N' products, it can be predicted that the number of products n_{12} that will fail between two values u_{k1} and u_{k2} of the service quantity is:

$$n_{12} = N'\left[F(u_{k2}) - F(u_{k1})\right]. \quad [2.8]$$

In the example above, it can therefore be predicted, thanks to this calculation and knowing only function F, that 137 products will fail in the service range [3–7] out of 1,000 products initially activated.

2.5.2. *Generic failure rate models*

In order to quantify reliability, the ability to perform analytical calculations based on the previous equations is required. For this purpose, generic analytical models are used to characterize the relationship that exists between the evolution of the failure rate and the service quantity. In practice, it can often be observed that this evolution follows the shape of a bathtub curve which consists of three successive stages:

1) At the onset of the product's life, the failure rate is high due to early life failures. For example, in Figure 2.9, early life failures stop as of 4 units of service. Such failures can be avoided by taking two suitable precautions during the design, manufacture, validation and operation stages. This is known as the burn-in process.

2) Then, the failure rate tends to level out. Failures occur randomly, which is convenient to apply reliability theory. For example, in Figure 2.9, this domain corresponds to a service quantity ranging between 4 and 14.

3) Lastly, the failure rate quickly increases due to fatigue and wear out failures. In Figure 2.9 for example such failures appear as of 14 units of service. Nevertheless, products which are not subject to wear and fatigue never reach this last stage, in particular signal processing electronics and software. As for products vulnerable to these phenomena, such as power transmission, maximum lifespan is estimated thanks to endurance and fatigue tests.

The two parameterized models most used for reliability calculations are described in the next section.

2.5.2.1. *Exponential probability of failure distribution*

This model is based on the assumptions that the failure rate does not depend on service quantity and that failures occur randomly throughout the life of the product:

$$h(u) = \lambda. \qquad [2.9]$$

As a consequence, the probability of failure distribution is exponential:

$$f(u) = \lambda e^{-\lambda u}. \qquad [2.10]$$

The main benefit of this model is that the probabilities of failure and success can be expressed in easy-to-manipulate mathematical forms:

$$F(u) = 1 - e^{-\lambda u} \qquad [2.11]$$

$$R(u) = e^{-\lambda u}. \qquad [2.12]$$

This form can be simplified if the failure rate is very low compared with service life:

$$F(u) \approx \lambda u \qquad [2.13]$$

$$R(u) \approx 1 - \lambda u. \qquad [2.14]$$

As previously explained, exponential probability distribution is well suited for products that are not subject to wear and fatigue phenomena. This model is also widely used in situations where this characteristic is not verified, even though technically it is not specific enough, because it makes reliability calculations much simpler.

2.5.2.2. *Weibull distribution*

The Weibull model introduces two additional parameters so as to model either early life effects or fatigue and wear out effects:

$$h(u) = \frac{b}{a}\left(\frac{u}{a}\right)^{b-1} \qquad [2.15]$$

$$f(u) = \frac{b}{a}\left(\frac{u}{a}\right)^{b-1} e^{-\left(\frac{u}{a}\right)^{b}} \qquad [2.16]$$

$$F(u) = 1 - e^{-\left(\frac{u}{a}\right)^{b}} \qquad [2.17]$$

$$R(u) = e^{-\left(\frac{u}{a}\right)^{b}}. \qquad [2.18]$$

Parameter b indicates the decrease ($b < 1$) or the increase ($b > 1$) of the failure rate as a function of service quantity. Parameter a indicates the rate at which this evolution occurs with respect to service quantity. If $b = 1$, the Weibull model is equivalent to the exponential model with $\lambda = 1/a$.

The Weibull model is mainly used when it is crucial to take into account the wear or fatigue effects but it rapidly leads to complex calculations.

2.5.3. *Reliability of element associations*

This paragraph introduces reliability calculations for associations of the most generic elements. Here, calculations are performed under the assumption that the faults of different channels are independent. Consequently:

– the occurrence probability of n events is the product of probabilities associated with each event; and

– the occurrence probability of one of the possible events is the sum of probabilities associated with each event.

2.5.3.1. *Elements associated in series*

The reliability of a system S that consists of n elements in series is given by:

$$R_S(u) = \prod_{i=1}^{n} R_i(u). \qquad [2.19]$$

For a system with two elements 1 and 2 whose failure rates are constant and respectively equal to λ_1 and λ_2, this probability can be reduced to:

$$R_S(u) = e^{-(\lambda_1+\lambda_2)u}. \qquad [2.20]$$

If failure rates are extremely low, then the probability of failure is approximately equal to the sum of failure rates 1 and 2.

2.5.3.2. *Elements associated in parallel*

Consider a system S consisting of n active and fail-safe channels associated in parallel. Assume the probability of failure of a channel does not depend on its contribution rate to the function (pure parallel association). The reliability of the system is then:

$$R_P(u) = 1 - \prod_{i=1}^{n} [1 - R_i(u)]. \qquad [2.21]$$

Regarding power systems, the failure rate of an element generally depends on its work rate, in other words its contribution to the implementation of the function. For instance, this is true for a two-actuator system in active/active mode, operated in force summation on a single load, as is often the case for flight controls. In the event that one of the system elements has a fault, other elements must increase their contribution, which in turn increases their failure rates for the same service quantity. This phenomenon does not appear in information systems, where the reliability of an element does not depend on the importance of its contribution to the function. Detailed calculations of the probability of failure must therefore be carried out carefully for systems whose elements are solicited differently depending on whether or not some elements have failed.

This phenomenon can be taken into account by considering a system with two channels in shared parallel redundancy. The two channels are identical and independent. Their failure rates are constant and equal to λ_1 when the load is equally shared between both channels and to λ_2 when the entire load is borne by a single channel which then assumes the entirety of the function by itself. The reliability of the system is then given by:

$$R_P(u) = e^{-2\lambda_1 u} + \frac{2\lambda_1}{2\lambda_1 - \lambda_2}\left(e^{-\lambda_2 u} - e^{-2\lambda_1 u}\right). \qquad [2.22]$$

If $\lambda_2 = 2\lambda_1$, meaning the failure rate is doubled if the contribution to the function is doubled, then:

$$R_P(u) = e^{-2\lambda_1 u} + 2\lambda_1 u \quad e^{-\lambda_2 u}. \qquad [2.23]$$

If $\lambda_2 = \lambda_1 = \lambda$, the contribution rate to the function does not impact the failure rate, as in information systems. Therefore, the result of equation [2.21] is obtained again:

$$R_P(u) = 2e^{-\lambda u} - e^{-2\lambda u}. \qquad [2.24]$$

2.5.3.3. *Elements associated through majority vote*

For a system S with a two-out-of-three majority vote (the system is operational if two out of the three elements in parallel are operational), with three identical elements m and a voter v, reliability is expressed as:

$$R_{VM}(u) = R_v(u)\left[R_m^3(u) + 3R_m^2(u)(1-R_m)\right]. \qquad [2.25]$$

2.5.3.4. *Elements associated through detection–correction*

For a redundant architecture of the detection–correction type, it is possible to simply express the probability of failure, assuming that failure rates are kept constant. Consider for this, a system with two independent elements associated in detection–correction mode. Only one element assumes the function, while the other is isolated and on standby. The constant failure rates are equal to λ_1 for normal channel 1 and to λ_2 for standby channel 2. The global failure rate of the diagnosis and switching functions is assumed constant and equal to λ_{DC}. The reliability of the system is therefore expressed as:

$$\text{If } \lambda_2 \neq \lambda_1, \text{ then } R_{DC}(u) = e^{-2\lambda_1 u} + e^{-\lambda_{DC}u}\frac{\lambda_1}{\lambda_1 - \lambda_2}\left(e^{-\lambda_1 u} - e^{-\lambda_2 u}\right) \qquad [2.26]$$

$$\text{If } \lambda_2 = \lambda_1 = \lambda, \text{ then } R_{DC}(u) = e^{-\lambda u}\left(1 + \lambda u \;\; e^{-\lambda_{DC}u}\right). \qquad [2.27]$$

2.5.3.5. *Examples*

By way of graphical illustration, the probability of failure of the different types of redundancy are compared in Figure 2.10 as a function of service u and under the following conditions:

– each channel has a constant failure rate equal to $\lambda = 2\ 10^{-6}/u$;

– the voter and the diagnosis and switching functions are assumed perfect; and

– concerning the shared parallel redundancy (power transmission), the failure rate of a channel is multiplied by two whenever it is the only active channel.

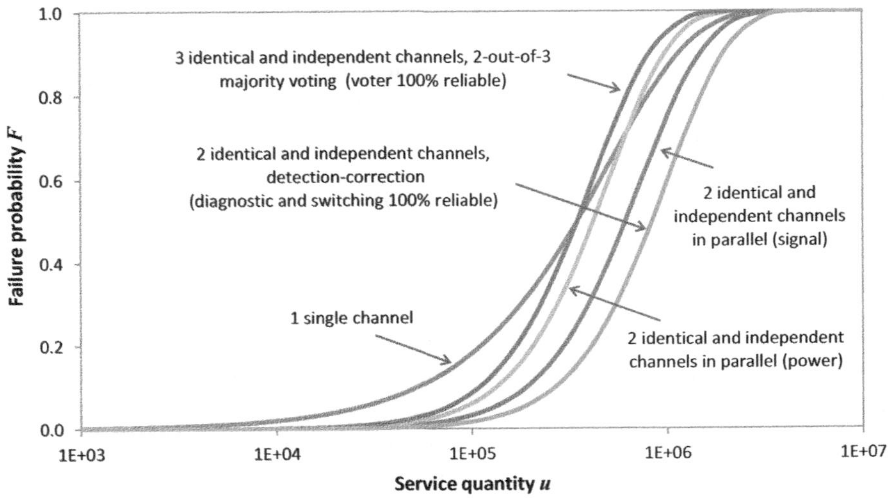

Figure 2.10. *Comparison of reliability depending on the type of element association. For a color version of the figure, see www.iste.co.uk/mare/aerospace1.zip*

The benefits for shared parallel redundancy of the detection–correction architecture and of the impact of the failure rate increase of the only active channel are obvious.

REMARKS.–

– Since simplified calculations rely on approximating $1-e^{-\lambda u}$ by λu, they quickly lead to false results, except for very low service quantities.

– When failure rates are low, in order to preserve the accuracy of calculations, it is best to work with probabilities of failure rather than with probabilities of success (reliabilities).

2.6. Short glossary of technical terms pertaining to reliability

Table 2.9 provides the meaning of the main words used in reliability engineering[5]. It is important to keep in mind that these terms may have a different meaning when used out of the context of technological system

5 A single asterisk indicates that definitions are adapted from [ISE 06]. A double asterisk indicates they are adapted from [TAY 76].

reliability. Sometimes the meaning of those words can even be reversed, as is the case for security and safety.

Availability*	Probability that a system or equipment operates in a satisfactory and efficient manner at all times.
Channel**	Entity consisting of its own specific elements.
Criticality	Indicator which combines the severity of a feared event, its occurrence frequency and how difficult it is to detect.
Danger	Something assigned with a feature that makes it a threat. Danger is the source of risk.
Dormant, latent, hidden fault	Fault whose effect is not detected during a given period of time.
Error	The result of an incorrect action (for example during the stages of specification, design, calculation, implementation or operation).
Failure mode	Failure mechanism or manner in which an element can fail.
Failure*	Expression of a fault leading to the interruption of the system's ability to perform a function under given operating conditions.
Fault*	Unauthorized deviation from a property or a characteristic of a system with respect to the normal/usual/standard result.
Gravity	Intensity measurement of consequences that may arise following the occurrence of an undesirable event (or of a random hazard).
Malfunction	Flickering irregularity of the implementation of a desired system function.
Redundancy**	Duplication of active channel(s) or implementation of backup channel(s) in order to increase reliability.
Reliability*	Ability of a system to perform a function during a certain amount of time, in a given environment and with a given objective.
Risk	Event to come, uncertain and that can potentially cause damages due to its dangerous nature.
Safety	Ability of a system to avoid generating danger for people, goods and the environment.
Security	Absence of threats posed by malice.
Severity	Direct consequence of a failure in terms of harm to people, to goods and to the environment.

Table 2.9. *Glossary of important technical terms relating to reliability*

3

Hydraulic Fluid and its Conditioning

3.1. Needs and constraints

3.1.1. *Opportunities and constraints in hydrostatic power transmission*

Within hydraulic systems, the energy transmission function is carried out via liquid matter transfer. This matter transfer, inextricably tied to energy transfer, provides the opportunities to perform two extra functions:

– *Transport heat.* Matter transfer is used to carry out a heat transfer function. The aim of this is to keep hydraulic components within their working temperature range to guarantee satisfactory performances and lifespan. For example, the hydraulic fluid is used to evacuate heat generated by energy losses caused by leakage and friction. It can also be used to heat up components which operate in low temperature environments.

– *Lubricate.* Liquid matter is used to lubricate contact areas between moving solid bodies within components. For example, this is useful between the spools and the sleeves of hydraulic valves or between the pistons and the piston chamber of hydraulic pumps where the relative velocity is in the meter per second range.

Unfortunately, all this also creates harsh constraints on the design, the integration and the operation of products relying on hydraulic power transmission.

On the one hand, the fluid has to be properly conditioned in order to correctly perform the three functions above, over the entire working range and during the entire life of the aircraft. It is therefore crucial that the fluid be clean and present in sufficient quantity. This requirement imposes strict segregation between hydraulic systems, in order to prevent leakage and pollution from spreading. Extra fluid volume is made available to enable the hydraulic fluid to dilate or retract depending on pressure and temperature. The physico-chemical characteristics of the fluid must remain in a range that is in accordance with the expected performance.

On the other hand, respect for the environment and for people is primordial and must be ensured. This can be achieved mainly by properly containing, collecting or re-directing possible external leakage.

Lastly, it is important to ensure safety, in particular by limiting fire hazards or the propagation of fire via the hydraulic fluid.

Therefore, the architecture of a hydraulic system must satisfy both lubrication requirements, and power and heat transfer requirements. It must also be based on a well-conditioned fluid, which must be resistant enough to fire, and be respectful of people and the environment.

3.1.2. *Actual hydraulic fluid*

In a hydraulic system, instantaneous and long-term performances are heavily impacted by the effective properties of the hydraulic fluid. In practice, actual hydraulic fluid is very different from pure hydraulic fluid. Indeed, on the one hand, it can contain gas produced by its partial vaporization when its pressure reaches vapor pressure[1]. On the other hand, it is affected by solid, liquid and gaseous pollutions (see Figure 3.1).

Part of the solid pollution is caused by a flawed initial cleaning process and by maintenance operations, while the rest results from in-service abrasion and erosion. Liquid pollution is mainly a consequence of the presence of air humidity, water with which the fluid can come in contact. Gaseous pollution can take the form of air pockets originating from a bad initial bleeding. It most often arises because of a liquid's ability to dissolve gases proportionally to pressure and fluid volume. This particularly applies

1 Vapor pressure of hydraulic oils is extremely low. For example, 66 Pa absolute for a type IV phosphate-ester at 93°C or 1000 Pa for a type V at 100°C.

to the air the liquid is in contact with. For example, hydraulic oils used in aeronautics can typically dissolve 5–10% of their volume in air at atmospheric pressure. When the pressure of the liquid drops below the saturation pressure, the liquid becomes oversaturated: dissolved gases transform into free gas bubbles. This is called the desorption[2] process. Conversely, when the pressure of the liquid rises above the saturation pressure, gas bubbles dissolve again in the liquid. Unfortunately, desorption is quasi-instantaneous whereas the dissolving process is much slower. This explains why free gas bubbles, of typical diameter lower than 0.1 mm, are always present in service. They form a gaseous pollution that is difficult to eliminate. When bubbles agglomerate at high points or in dead zones of the hydraulic circuit, they form free gas pockets.

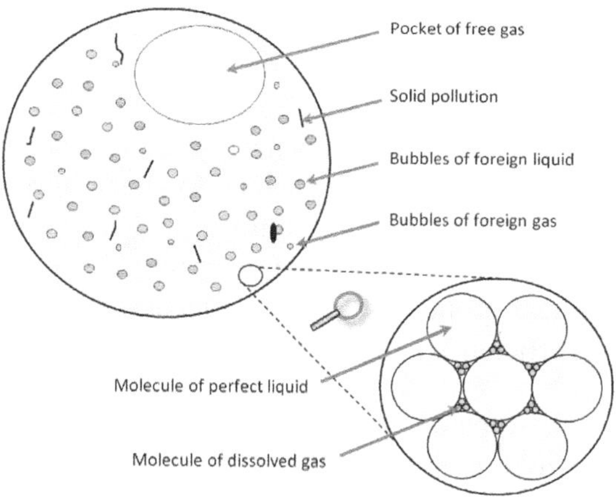

Figure 3.1. *Actual fluid and its pollution*

Clearly, solid pollution can lead to jamming between two moving parts when efforts to initiate motion are insufficient (for example between the spools and the sleeve of a servovalve). However, if these efforts are sufficient (for example at the level of a hydraulic pump valve plate), solid pollution deteriorates the system, slowly at first, but it generally quickly leads to a loss of function by jamming or rupture.

The presence of gas in the hydraulic fluid has an immediate impact on performances. This is because gas very sharply increases compliance of the

2 It is worth noticing that, as the pressure of a hydraulic fluid decreases, gases are released well before vaporization. However, this is generally not the case for fuel.

hydraulic fluid which, in turn, directly affects the stability and dynamics of hydraulic power users (for example, flight control actuators or brakes).

As for commercial aircraft hydraulic fluids, the presence of water and oxygen combined with a high, and often excessive, operating temperature triggers hydrolysis, oxidation and pyrolysis chemical reactions [BEH 10]. These reactions produce acid which attacks materials (metals, paint, skin, etc.) and which forms deposits on walls in the long run. In service, a maximum of 0.8% of water is tolerated. Deposits and corrosion quickly form above 3% of water, which can cause flight control actuators to operate erratically.

This explains the importance of filtration, bleeding and health status monitoring. These concepts will be addressed further in the chapter.

3.1.3. *Physical properties*

From the standpoint of physical behavior, the actual hydraulic fluid is subject to various effects: inertia, resistance to shear (viscosity) and elastic deformation under pressure (compressibility) or due to temperature change (thermal dilation). These effects are respectively quantified by four parameters or physical properties: specific density, viscosity, bulk modulus (compressibility coefficient) and dilation coefficient.

Unfortunately, these parameters significantly depend on the temperature and pressure of the hydraulic fluid. This complicates the designers' task and imposes operational constraints. Table 3.1 illustrates the variations of these parameters for an aerospace phosphate-ester fluid.

	From 0 to 100°C at 0 bar gauge	From 0 to 207 bar gauge at 50°C
Density	-7.5%	?
Bulk modulus (isothermal secant)	-44%	+11%
Kinematic viscosity	Divided by 10	+57%

Table 3.1. *Typical variations of the phosphate-ester hydraulic fluid with pressure and temperature [SAE 08]*

Fluid properties and their variation have a major impact on hydraulic systems. Therefore, it is especially important to consider the following aspects.

3.1.3.1. *Fire-proofing*

In the field of aeronautics, the fluids used are fire-proof. For military or commercial applications, the oils are hydrocarbon-based synthetic oils (for example, according to standard Mil-H 83282 [MIL 86] typically with an auto-ignition temperature of 354°C and a heat of combustion of 9.1 kcal/g). As for commercial applications, fire-proofing is improved thanks to the use of phosphate-ester type synthetic oils (for example according to standard AS1241 [SAE 97] typically with an auto-ignition temperature increased by 48% and a heat of combustion reduced by 28%).

3.1.3.2. *Insufficient awareness of effective physical properties*

Little is known about the sensitivity of certain physical properties of the fluid to temperature and pressure because it is difficult to measure. Therefore, it is unfortunately only partially described and documented by hydraulic oil suppliers.

3.1.3.3. *Impact of compressibility on performances*

Between -50 and +100°C, the volume of a given mass of aerospace hydraulic fluid typically gets increased by 12% due to dilation. Between 0 and 207 bar, the volume gets reduced by 1.5% due to the compressibility effect. Although this fluctuation of specific density with pressure is small, it occurs very quickly. As a result of this phenomenon, the compressibility effect can be the major cause responsible for the limitation of dynamic performances of actuators.

3.1.3.4. *Very high sensitivity of viscosity to temperature*

Viscosity is extremely sensitive to temperature. It is typically divided by 10 between 40 and 0°C, and it is again divided by 10 between 0 and 100°C. Assuming that in very small gaps between moving parts, viscous forces are proportional to viscosity and that leakage flow rate is inversely proportional to viscosity, then the importance of maintaining the fluid within an acceptable temperature range is understandable. At very low temperatures, a rise in viscosity increases viscous friction forces and causes mechanical efficiency to drop. On the contrary, at high temperatures, a decrease in viscosity increases leakage in gaps and deteriorates volumetric efficiency.

3.1.3.5. *Impact of free gas*

The presence of free gas significantly impacts the effective Bulk modulus which quickly deteriorates: typically, the presence of 1% volume of free gas in the hydraulic fluid at 200 bar makes it twice as elastic. As long as it remains within the acceptable range, the proportion of free gas has little effect on the other physical characteristics.

Hence, the formulation of aeronautic hydraulic fluids is mainly driven by their fire-proof property, their physico-chemical stability and the minimization of viscosity variation as a function of temperature.

3.2. Fluid conditioning

Table 3.2 provides a summary of requirements and their corresponding solutions implemented to ensure the fluid is properly conditioned. Corresponding architectural solutions are detailed in the following sections.

Conditioning requirement	Solution implemented
There is a sufficient quantity of fluid and it has sufficient space	Centralized reservoir, local compensators
The fluid is clean	Bleeding (releases gas pockets) Separation (of free gases and exogenous liquids) Filtration (isolates solid pollution) Pressurization (prevents the appearance of gaseous pollution)
The fluid is at the correct temperature	Heating (natural or forced) Cooling (natural or forced)
The fluid is not dispersed in the environment	Collection of external leakage

Table 3.2. *Requirements and generic solutions implemented to condition the fluid*

3.2.1. *Fluid in sufficient quantity*

The architecture of a hydraulic system must guarantee the presence of hydraulic fluid at all times, particular at the pump suction stage. It is therefore primordial to satisfy various needs.

3.2.1.1. *Facilitate pump suction*

The presence of fluid must always be ensured, including in the presence of loading caused by sideways or negative accelerations.

3.2.1.2. *Allow volume to vary as a function of pressure and temperature*

The volume of hydraulic fluid contained in the circuit and equipment varies depending on temperature and pressure. For example, for a large passenger jet of the Boeing B747 type, the total hydraulic fluid volume, i.e. more than 650 l, typically varies by 50 l between -40 and +60°C.

3.2.1.3. *Compensate for the volume variation caused by equipment*

The volume of fluid contained in the equipment varies as a function of its configuration. For example, a single-rod type cylinder for landing gear extension/retraction contains less fluid when the rod is pushed in all the way than when it is pulled out. A second example is the high pressure hydraulic accumulator: it contains less fluid in the absence of hydraulic pressure when the airplane is at a standstill. Figure 3.2 illustrates this effect using the relative variation of fluid volume in the reservoir as a function of the activation of hydraulic loads during a flight.

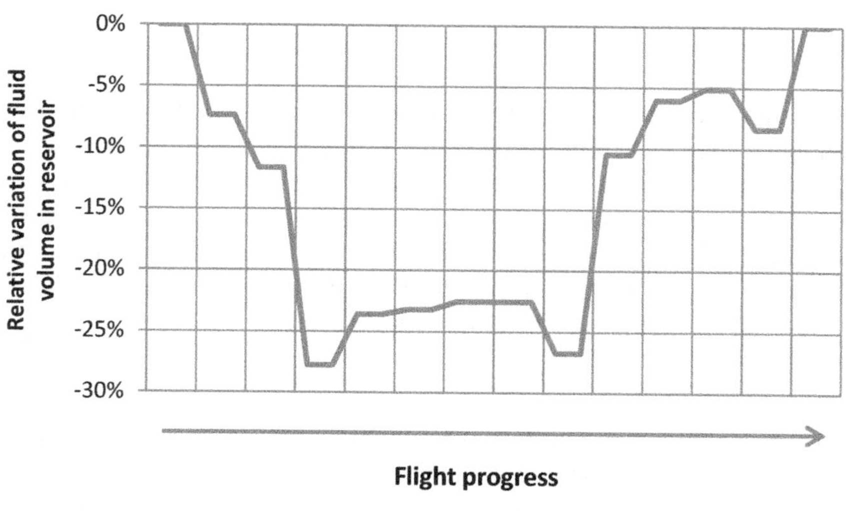

Figure 3.2. *Example of fluid volume variation in the reservoir of a hydraulic circuit caused by geometric effects*

3.2.1.4. *Compensate for external leakage*

Most often, external leakage is a consequence of normal operation or of disassemblies performed during maintenance. For instance, a maintenance operation on a servo control must be initiated only if leakage from the rod is more significant than 1 drop per 25 full stroke cycles (corresponds to the cumulated stroke of a 2-hour flight). Across the entire hydraulic equipment of a Boeing B747, leakage ranging from 5 to 8 drops is tolerated per minute, with a maximum total of 100 drops (5 cm^3) per minute for the entire airplane.

The entire set of functions listed above are in general carried out at the reservoir. The purpose of the reservoir is to make it possible to adjust the volume of fluid in the pipes. In a single-aisle aircraft, the total reservoir volume typically represents 15–25% of the total hydraulic fluid volume at 20°C. This is because most of the fluid is contained in hydraulic lines and equipment. This explains why the reservoir volume is so small compared to the volume of industrial set-ups, for which the volume is generally 4–10 times the volume pumped out by the pump in one minute (this standard engineering practice has the added benefit of giving free gas bubbles enough time to surface within the reservoir).

3.2.2. Pressurization and charging

Pressurization and charging are used to *push back* the occurrence of pollution in free gas form, in particular at the pump suction stage. Hence, there is less risk that the fluid pressure in the system will locally drop below saturation pressure, vapor pressure or simply atmospheric pressure. Concerning this latter risk, it is therefore ensured that seals toward ambient air always work with the same differential pressure direction, the low pressure side being the side in contact with the air.

3.2.2.1. *Pressurization*

The purpose of pressurization is to impose a minimal pressure in the low pressure area of a hydraulic system (typically, on commercial aircraft, pressurization is carried out at 4.5 bar absolute and must be maintained at least 12 hours after engines are switched off). However, given that many aircraft do not use hydraulics for critical functions, as is the case for regional turbo-props for example, they do not rely on pressurization for simplicity. Regarding other aircraft, pressurization is carried out at the reservoir following one of three existing generic approaches:

1) Air pressurized reservoir. The hydraulic fluid is pressurized by the air originating from engine bleeding, with no physical separation between the air and the hydraulic fluid (free surface pressurized reservoir). This is a common solution on commercial aircrafts because it facilitates the separation of free gases by centrifuging the input fluid and then letting gravity lead bubbles to the surface. However, this solution requires particular construction features, such as the addition of anti-g walls to avoid sloshing under sideways acceleration, or walls enabling the essential hydraulic circuit to be fed in priority. Furthermore, this solution has the disadvantage of introducing a coupling between hydraulic systems and air systems. The air used for pressurization which is in direct contact with the hydraulic fluid must be filtered, decanted and its pressure must be controlled. Figure 3.3 details the functions implemented to pressurize the reservoirs on a single-aisle commercial aircraft. A picture of the air pressurized reservoir of Airbus A380 is shown on the left in Figure 3.5.

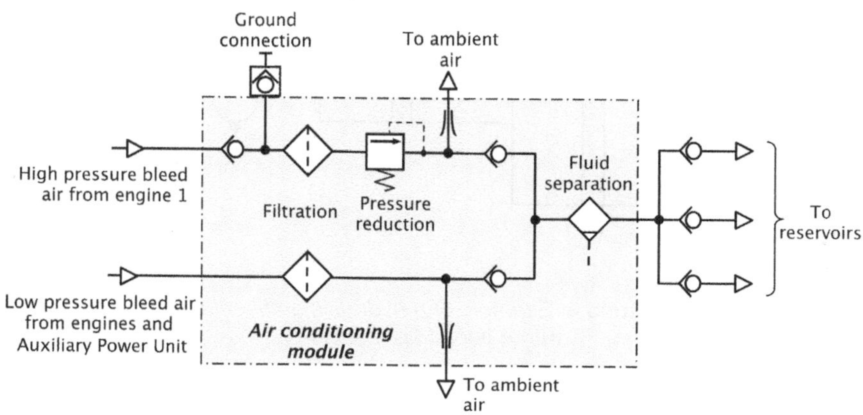

Figure 3.3. *Example of pneumatic pressurization system for the hydraulic circuits of a commercial airplane [DIE 14]*

2) Bootstrap reservoir. The hydraulic fluid is pressurized by a piston acting under the pressure supplied by hydraulic pumps. Since there is only

hydraulic fluid in the reservoir, this solution is tolerant to load factors and it is therefore well suited for fighter aircrafts or even helicopters. Although it removes coupling with air systems, this solution requires additional devices to maintain pressurization at the end of the flight when pumps are switched off, as well as bleeding devices (see Figure 3.4). On newer aircrafts such as Boeing B787 and Airbus A350, the introduction of automatic bleed valves [BAR 09] made it possible to overcome the need for manual interventions, such as those that were required on the Douglas DC10 or on the Mc Donnell MD80 [MAD 84]. This makes it possible to bypass the reservoir access constraint. Moreover, environmental-friendly bottles are installed to collect fluid at the bleed valves, drains and low pressure relief valves. The right picture in Figure 3.5 shows the bootstrap reservoir of the military transport aircraft Casa C295.

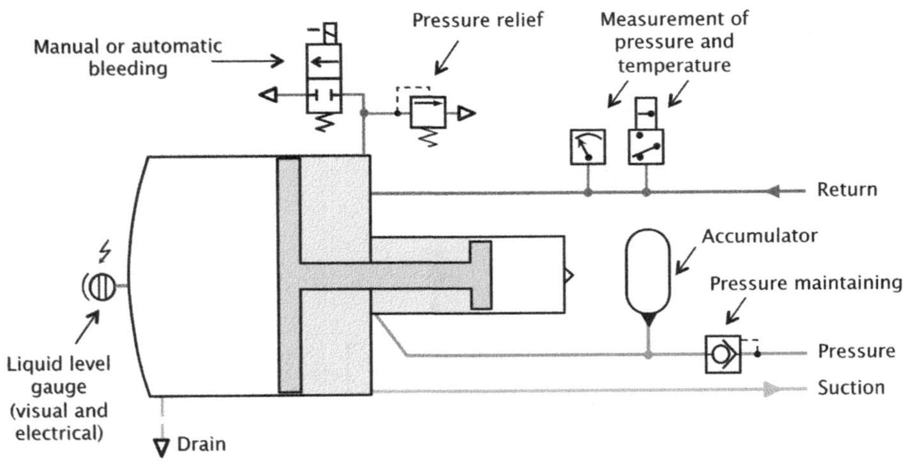

Figure 3.4. *Example of a hydraulic architecture with a bootstrap reservoir*

3) The hydraulic fluid is pressurized by mechanical pre-loading (for example, a piston and spring compensator). This solution is used in sub-systems or equipment that are isolated from the centralized hydraulic system during certain phases of flight (for example, landing gears) or depending on their operating mode (for example, flight control actuators in passive mode).

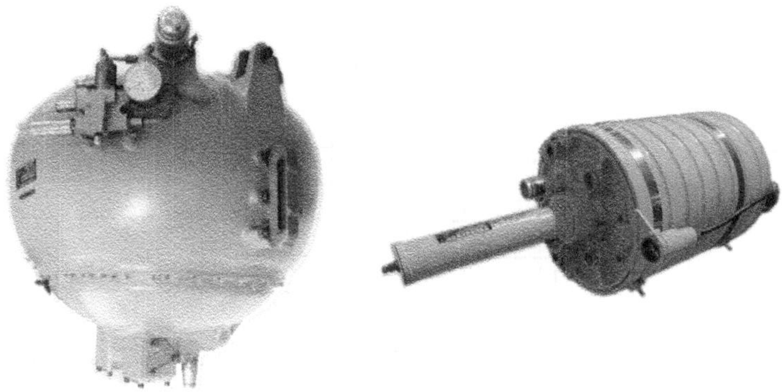

Figure 3.5. *Air pressurized reservoir of the Airbus A380 on the left, and bootstrap reservoir of the CASA C295 on the right*

3.2.2.2. *Charging*

Charging consists of force-feeding fluid (for example in a cylinder chamber) in order to maintain fluid at a minimum pressure. In hydraulic pumps, charging is carried out by a dedicated pump: the feed pump. As for actuating equipment, it often has a charging (re-feeding, anti-cavitation) function. Indeed, they can be led to operate as hydraulic generators when the load is a driving load. These conditions are met, for example, when speed reversal occurs as a result of the inertia of the driven load, or due to the application of external forces (e.g. aerodynamic forces, driving forces supplied by the active channel in active/standby mode or parking push back forces for nose landing gear steering servo controls). Architectural solutions implemented to satisfy this need will be described in Chapter 6.

3.2.3. *Filtration*

Filtration consists of isolating pollution from the hydraulic fluid. In aerospace, solid pollution is quantified according to standards (NAS 1638, SAE-AS4059 RevE or NAV AIR 10-1A-17) which define contamination classes and grades them. For example, the NAS 1638 standard, widely used in commercial aerospace, defines for each class the acceptable maximum number of solid particles in 100 ml of hydraulic fluid and for five particle size ranges, as shown in Table 3.3. It can be noted that:

– moving up one class is equivalent to tolerating twice as much contaminants for each size range;

– for a given class, the number of tolerated contaminants is typically divided by 5 when moving up one size range.

Particle size (µm)		5 to 15	15 to 25	25 to 50	50 to 100	> 100
NAS 1638 grade: number of particles in 100 ml sample of hydraulic fluid	00	125	22	4	1	0
	0	250	44	8	2	0
	1	500	89	16	3	1
	2	1000	178	32	6	1
	3	2000	356	63	11	2
	4	4000	712	126	22	4
	5	8000	1425	253	45	8
	6	16000	2850	506	90	16
	7	32000	5700	1012	180	32
	8	64000	11400	2025	360	64
	9	128000	22800	4050	720	128
	10	256000	45600	8100	1440	256
	11	512000	91200	16200	2880	512
	12	1024000	182400	32400	5760	1024

Table 3.3. *Maximum number of solid particles tolerated in 100 ml of hydraulic fluid as a function of their size and contamination grade*

Filtration is generally performed "in series", meaning filters are added in series on the path of hydraulic power[3]. This configuration is shown in Figure 3.6. In series, filtration is a pretty simple process. It requires few components other than filters. However, its efficiency directly depends on the flow rate of the fluid flowing in the network on which filters are located. This is why it is necessary to externally drain components that are expected to function on low flow rates for long periods of time (for example, main pumps). The fluid output of the drain is also filtered and returned to the reservoir with as little resistance as possible in order to avoid raising the pressure in the body of the element drained. Furthermore, filters can only be fit on hydraulic lines in which the direction of the flow cannot be functionally reversed (otherwise isolated contaminants are dragged away with the flow). Depending on applications, filtration involves different types of filters.

3 In order to break free from this dependence, there also exists, in industrial applications, a "parallel" filtration solution. It involves an additional hydraulic circuit, specifically designed to suction the fluid from the reservoir, to filter it and return it to the reservoir.

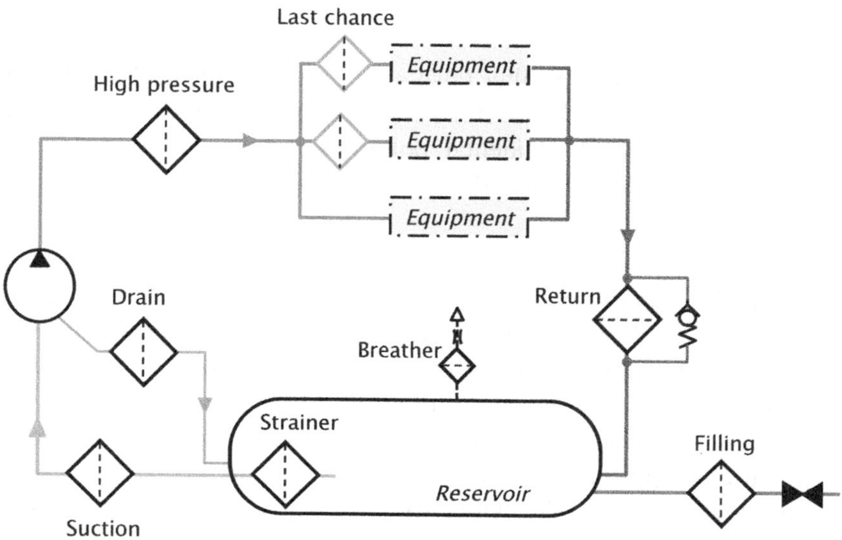

Figure 3.6. *In series filtration architecture*

3.2.3.1. *Suction filter*

Concerning industrial applications, a strainer and/or a suction filter are responsible for protecting the pump against solid pollution coming from the reservoir. In order to avoid vaporization or desorption of the fluid which is at low pressure, these filters should not oppose too much hydraulic resistance. This is why their absolute filtration fineness (i.e. the size of the particles that are 100% filtered out) is rather high: typically, about 50 µm. These filters are hardly ever present on the hydraulic systems of aircraft.

3.2.3.2. *Drain filter*

A filter is fitted on drain lines in order to avoid polluting the reservoir. These filters typically have an absolute fineness ranging from 15 to 25 µm.

3.2.3.3. *High pressure filter*

One or more filters are fit on high pressure lines to protect the network and the users of hydraulic power. Since the fluid is at high pressure and since certain users are very sensitive to pollution (for example servovalves), a filtration quite fine is implemented even though it leads to the loss of several tenths of a bar in loading. High pressure filters typically have an absolute fineness ranging from 15 µm.

3.2.3.4. *Integral filter*

A high pressure internal filter (also called the "last chance" or "integral" filter) is incorporated in the most sensitive equipment, such as servovalves. Its purpose is to avoid accidental pollution resulting from maintenance operations. Integral filters typically have an absolute fineness of 3 µm.

3.2.3.5. *Return filter*

A return filter is fit on the low pressure line to avoid transmitting to the reservoir the solid pollution generated by hydraulic power users or maintenance operations. Thanks to this filter, particles of size typically higher than 3–5 µm are isolated. Filter clogging leads to an excessive return pressure which can cause damage to hydraulic components located on the low pressure line upstream the filter. In order to avoid this, a pressure relief valve is mounted in parallel with this filter and it opens when there is a differential pressure of a few bars. On certain hydraulic systems, in particular those with an unpressurized reservoir, the return filter is also equipped with a re-feeding valve. This helps prevent the appearance of free gas in the line downstream the filter during transient pressure drops.

3.2.3.6. *Breather filter*

When reservoirs are open to the air, a breather filter is used to prevent the introduction of polluted air in the reservoir when the fluid level drops. Breather filters are characterized by an absolute fineness of 10 µm.

3.2.3.7. *Filling filter*

The filling filter of the reservoir prevents the introduction of external pollution during refilling operations of the reservoir. Its absolute filtration fineness is typically of 15 µm.

Bleed valves, manual or automatic, are also installed to release free gas pockets during integration of after maintenance has been performed. The monitoring process of filter clogging is addressed in section 3.3.2.

3.2.4. *Thermal management*

Various methods can be adopted to maintain the hydraulic fluid within its working temperature range and not let it become too hot or too cold. For example, Figure 3.7 shows that, on board a commercial airplane, hydraulic equipment must ensure nominal performance level when the fluid temperature is higher than or equal to -15°C. The maximum temperature tolerated for the hydraulic fluid is 95°C and the minimum optimal temperature for the fluid is 20°C.

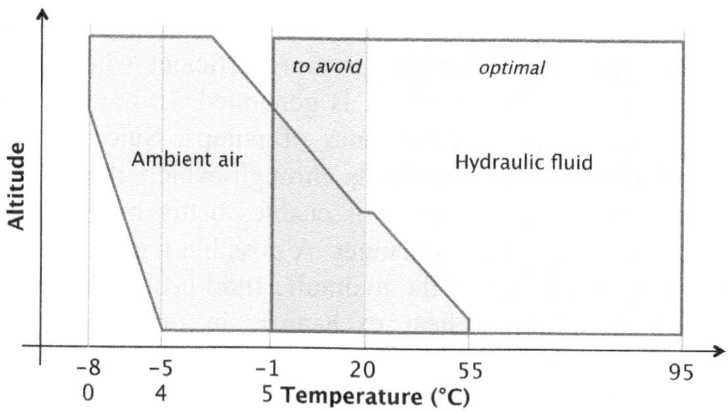

Figure 3.7. *Example of a hydraulic fluid temperature range as a function of altitude for a commercial airplane [BEH 13]*

3.2.4.1. *Cooling*

The hydraulic fluid helps evacuate most of the heat generated by energy losses. These losses have both beneficial and parasitic effects: they are functional at the level of power control valves but they are parasitic because of internal leakage, friction and pressure drops in components and in hydraulic lines. Therefore, fluid heating has to be controlled in order to maintain its viscosity within its useful range compatible with performances, and to limit its chemical deterioration. The following simple calculation is a good illustration of the need for cooling. On a commercial airplane, it is observed that in cruise conditions, the cruise internal leakage flow rate from equipment typically represents 5–20% of the flow rate capacity of hydraulic pumps. Furthermore, power lost in one of the main pumps typically represents 10% of its hydraulic nominal power. In the end, without supplying any actuating power, between 15 and 30% of the rated driving mechanical power of pumps[4] has to be spent on simply maintaining equipment at its nominal supply pressure.

4 For example, when one of the main hydraulic pumps of Boeing B777 delivers 20% of its nominal power, its global efficiency is of the order of 57%. Hence, power dissipated in the pump then exceeds 9.5 kW.

The primary heat sink is the surrounding air with which components and hydraulic lines can exchange heat by convection. However, this method is generally not powerful enough. It is therefore necessary to cool the hydraulic fluid by forcing further heat exchanges.

Similar to filtration, cooling is implemented in series[5] by fitting heat exchangers on return lines. Cooling is more efficient when performed as closely as possible to where the heat is generated. In practice, this means carrying out cooling on the drain lines of pumps. Since these lines are subjected to low pressures, the walls through which the heat exchange occurs can then be made thinner. This enables better heat conduction and lowers the weight of the heat exchanger. A possible method to achieve this consists of taking advantage of the hydraulic fluid heat to heat up fuel with the help of a liquid-liquid heat exchanger, in order to improve fuel combustion. This solution, although widely used, introduces coupling between the hydraulic and fuel systems and this can potentially cause safety issues (for example in case one of the exchanger walls ruptures). Otherwise, the fluid can be cooled by the ambient air thanks to air-liquid exchangers. However, this alternative solution, introduces coupling with aerodynamics because of the drag generated by the air intake.

These solutions can be combined, as is the case with Airbus A380. In the normal mode, fluid originating from the drains of the main pumps is cooled down by air-oil exchangers. External air flow can be forced by hydraulic motors. Exchangers are fit in the flap-track fairing, as shown in Figure 3.8. In the backup mode, cooling is assumed by oil-fuel exchangers. It is worth noting that the main pumps of Airbus A380, which deliver more than 150 l/min at 350 bar (i.e. typically 90 kW), are equipped with a pump specifically dedicated to reinforcing the drain flow rate.

3.2.4.2. *Heating*

The hydraulic fluid must be heated during the ground starting stage or when the surrounding temperature is very low during a flight. On the one hand, the goal is to ensure that hydraulic equipment operates properly, for example to prevent excessive lowering of the gains of servo control metering valves. On the other hand, it is necessary to limit distribution of pressure drops between generators and the most remote users (for example ailerons

---

5 In certain industrial applications, cooling can be performed by the parallel filtration circuit which then needs to be fitted with a heat exchanger.

and control surfaces located in the tail of aircraft or the nose landing gear). Indeed, the return pressure cannot be too high at the level of the equipment and the supply pressure must be compatible with the minimal power and dynamic performances. It should also be noted that for air pressurized reservoirs, certain elements of the air circuit (for example check valves) must be heated in order to keep them from freezing shut.

Figure 3.8. *Air/oil cooling on Airbus A380*

Heating is partly carried out naturally thanks to permanent internal leakage of components such as manifolds, servo controls and pumps. The simplest complementary solution consists of heating the fluid by converting its pressure energy into thermal energy while the fluid is flowing through hydraulic restrictions: if all the pressure energy is converted into heat transmitted to the fluid, a hydraulic fluid that goes from 207 to 0 bar and typically gets heated by more than 10°C. For the most remote users, internal leakage can be triggered on purpose by inserting a fixed orifice between the pressure and return lines. This solution has been used for a very long time on commercial aircraft equipment that can be located tens of meters away from generators, like the hydromechanical steering servo controls of rudders for example. This solution is simple; however, it creates a real dilemma. On the one hand, the leak needs to be significant enough at low temperatures to allow the fluid to be heated. This minimizes distribution losses and makes it possible for equipment to function properly. On the other hand, it needs to be weak enough to avoid wasting energy unnecessarily when the temperature of the fluid is high enough.

Recently, a more refined thermal management method was introduced on Airbus A350XWB to solve this dilemma [BEH 13]. First, it aims to increase the minimum temperature of the fluid at the level of users. In addition to having a beneficial impact on their efficiency, this makes it possible to reduce the diameter of hydraulic supply network pipes. Hence, as long as the flow is laminar in a pipe, the pressure drop decreases if the fluid temperature rises. Under a forced pressure drop, it is therefore possible to make the diameter of pipes smaller. Second, permanent leakage is progressively reduced when fluid temperature rises, until it gets completely canceled when temperature reaches 30°C: no pressure energy is wasted if heating is not needed. This self-regulating heating function was implemented simply and in a passive manner with the help of shape-memory materials. Unfortunately, the possible reduction of pipe diameter is limited. Indeed, a smaller diameter increases the hydraulic inertia of the fluid in pipes (see section 3.4.3), and this encourages the emergence of detrimental dynamic effects (hydraulic shock, pressure drops under transient flows, decrease of the natural frequency of hydraulic lines).

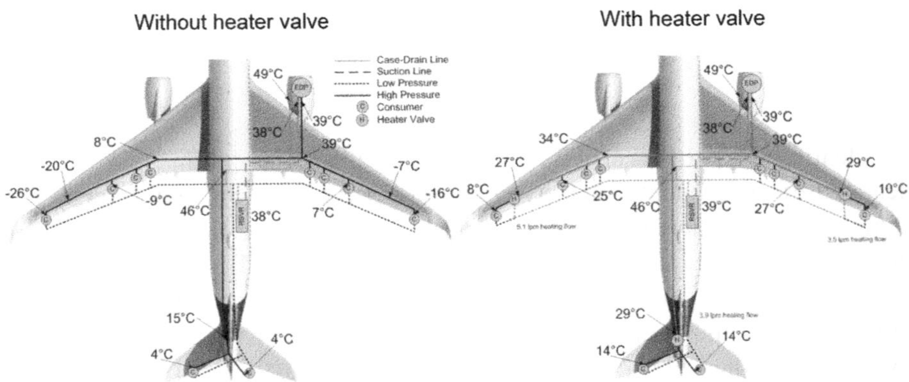

Figure 3.9. *Simulation of hydraulic fluid temperatures in an Airbus A350 cruising at -80°C ambient with and without adjustable heater valves [BEH 13]*

The case of the American space shuttle is interesting from the standpoint of hydraulic fluid thermal management [DAV 75]. Depending on the phase the mission is in, it is either necessary to heat up or cool down the hydraulic fluid Mil-H-83282 in order to maintain its temperature in the working range of -17 to +98°C. Heating is required before launch because hydraulic systems are close to cryogenic systems. It is also required during the orbital phase (which lasts between 7 and 30 days) because the ambient temperature is extremely low. In order to achieve this, the fluid is set in motion in the hydraulic circuit by dedicated circulation pumps. It is heated by a Freon-31 refrigeration loop that uses electronic systems as a heat source. In areas where initiating flow is impossible, electric heaters are fit on hydraulic lines. Additionally, cooling is required during other phases of the flight, such as re-entry in the atmosphere or switching off after landing. The hydraulic fluid is then set in motion by the same circulation pumps and it is cooled down by water spray exchangers.

3.2.5. *External leakage collection*

The hydraulic circuits of aircraft require hydraulic rotating machines such as the pumps of electric pumps, the motors of slats, flaps and backup power generators and power transfer units. The rotary dynamic seals of these hydraulic machines are subjected to significant slip speeds. Unfortunately, in order to reach the required lifespan, minimum external leakage has to be tolerated (for example 2 drops per service hour). This builds up to a non-negligible amount over the lifespan of the aircraft, for example 12 l in total for 120,000 flight hours. Hence, it is not acceptable to leave this external leakage spread freely outside the hydraulic system. The fluid is therefore collected by connecting the drains of the rotary seals (and the outputs of the pressure relief valves of the main reservoirs) to an external leakage collection bottle.

3.3. Monitoring and maintaining the fluid in working conditions

The hydraulic system is equipped with observation (monitoring) and action (maintenance) means. Their goal is to ensure the fluid is in good condition, meaning present in sufficient quantity, clean, at an acceptable temperature and properly pressurized. The display of monitoring information and the implementation of corrective measures are carried out via human-machine interfaces:

– directly at the level of the measurement spot;

– at the level of hydraulic ground service panels;

– at the level of the cockpit, for example for the Electronic Centralized Aircraft Monitor (ECAM).

These functions can be assisted or partially automated. This is the case on certain Airbus aircraft, where they are assisted by a Hydraulic System Monitoring Unit (HSMU). The observation is performed by sensors which output visual or electrical information of the proportional or on/off type.

3.3.1. *Fluid quantity*

The quantity of fluid can be measured by:

– a visual liquid level gauge located on the reservoir;

– a liquid level transmitter which outputs proportional electrical information and sends it to the cockpit and the ground service panel where it is displayed;

– a liquid level switch which transmits to the cockpit when the level drops below a given threshold.

The liquid level is typically corrected by adding fluid through the ground service panel: either directly from an external pressurized source or from an external barrel with the help of a hand-pump integrated in the service panel. It is worth noting that the reservoir is equipped with a drain valve which makes it possible to reduce the level of fluid if necessary.

3.3.2. *Cleanliness*

The cleanliness of the fluid is first of all measured by observing filter clogging. For this purpose, filters are equipped with a pop-up clogging indicator which is triggered when the filter causes an excessive pressure drop. These indicators are generally designed to be inhibited at very low temperature in order to avoid false readings when the fluid is very viscous. On the newest aircraft, clogging is indicated electrically and no longer requires visual inspection of filters. In the event of clogging, the corrective measure consists of replacing the concerned filters, if appropriate, after having identified and removed the pollution source. It is also worth noting that a fluid sampling set-up makes it possible to collect fluid samples that are analyzed in the lab in order to find out the level and the nature of the pollution present (typically by counting and identifying the nature of solid particles, by measuring the quantity of water, etc.).

3.3.3. *Pressurization – depressurization*

Pressurization monitoring is performed simply by a pressure gauge fit directly on the pressurization manifold of the reservoir. As for insufficient pressurization, it is detected by a pressure switch (for example with a threshold set a 2.5 bar absolute) and sent to the cockpit for processing and displaying.

If a maintenance operation involves opening the hydraulic circuit (for example if replacing a component), it is crucial to depressurize the concerned circuit beforehand in order to prevent leakage to the surroundings. Depressurization is performed by connecting the reservoir to ambient air. In general, this is done at the ground service panel. After the maintenance operation has been completed, pressurization is carried out by connecting an external air pressure source, for example connected to the pressurization air conditioning manifold.

3.3.4. *Examples*

Figure 3.10 completes Figure 3.6 regarding monitoring and maintaining in working conditions.

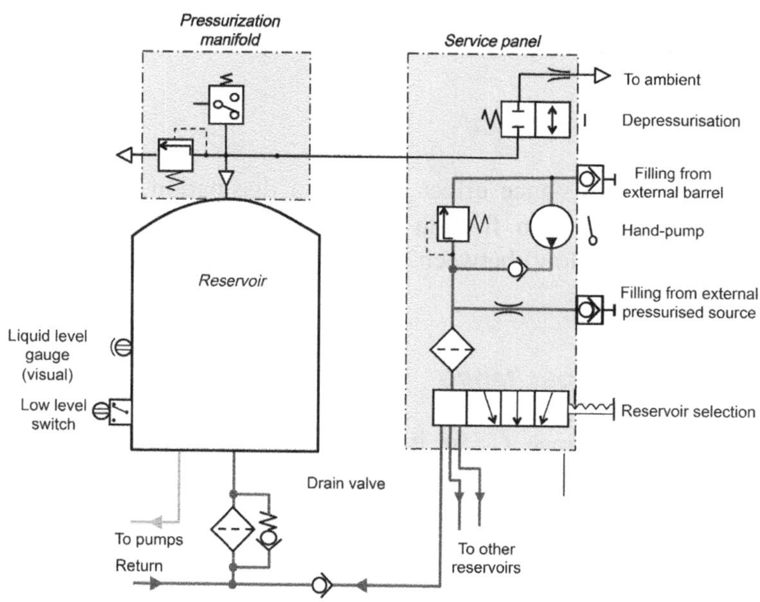

Figure 3.10. *Example of hydraulic architecture associated to the functions of monitoring and maintaining working condition for fluid conditioning*

Figure 3.11 shows the ground service panel of the green circuit of Airbus A350 which groups together the main functions previously introduced.

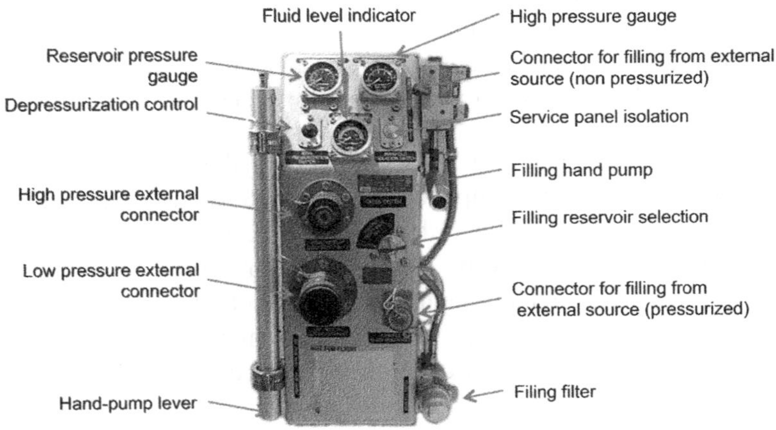

Figure 3.11. *Service panel of the green circuit of Airbus A350*

3.4. Energy phenomena caused by the fluid

The hydraulic fluid gives rise to the three basic energy phenomena [MAR02a]: the resistive effect, the capacitive effect and the inertia effect.

3.4.1. *Hydraulic resistance*

The hydraulic resistance effect leads to a dissipation of energy in heat form that originates from friction (in laminar flow) and partially elastic shocks (in turbulent flow) between particles. It will be described in detail in section 5.4.

3.4.2. *Hydraulic capacitance*

Hydraulic capacitance C_h, if it is the only capacitance present, is an energy conservation phenomenon which is caused by the compliance of the hydraulic fluid. It characterizes the relationship between infinitesimal increases in volume and in pressure in a given volume of fluid:

$$dV = \frac{V}{B} dP = C_h \, dP \tag{3.1}$$

Assuming that C_h is constant, the following relation is obtained after differentiating with respect to time:

$$Q = C_h \frac{dP}{dt} \qquad [3.2]$$

Therefore, in order to increase pressure P proportionally to time, it is necessary to inject in this volume a constant flow rate Q. This expression has the same form as the expression for the electrical capacitance which relates current I to the derivative of voltage U.

Hydraulic capacity can be expressed as a function of the fluid Bulk modulus B and the volume V of the considered fluid domain as follows:

$$C_h = V / B \qquad [3.3]$$

Unlike electrical capacity, hydraulic capacity can never be constant for a number of reasons:

– The volume V varies as a function of pressure since the fluid is compressible. It also varies because of the compliance of walls, especially that of flexible hydraulic lines. Lastly, it also varies due to the displacement of the moving parts of components, just as for example the volume in a cylinder chamber varies as a function of how far the rod is pulled out.

– The Bulk modulus B of the pure fluid varies as a function of temperature and pressure. For a type IV phosphate-ester at 25°C, B decreases by 25% when pressure drops from 550 to 207 bar, according to [SAE 08]. At 207 bar, it drops by a further 40% between 25 and 110°C. Additionally, it is impacted by the presence of free gas. Generally speaking, it is noteworthy that the presence of 1% volume of free gas at 200 bar divides the Bulk modulus by 2, which is heavily penalizing.

In light of this sensitivity, there are several ways of defining the numerical value of the hydraulic capacitance (isothermal, adiabatic or according to the speed of sound in the fluid). The fact that the hydraulic capacitance fluctuates makes the design task more difficult, especially when the robustness of performances has to be ensured. If there is a need for a mean value, an apparent Bulk modulus B_a is often defined. B_a accounts for the deformation effect of walls and for the presence of free gas. Typically, B_a is given a value between 7,000 and 8,000 bar for a fluid for 207 bar in rigid pipes and equipment.

The hydraulic capacitance also has an effect in the mechanical domain through hydromechanical transformers: it causes a hydromechanical stiffness effect K_{hm}. In the general case of an asymmetrical cylinder of hydrostatic section areas S_1 and S_2, hydromechanical stiffness is given by:

$$K_{hm} = B_a\left(\frac{S_1^2}{V_1} + \frac{S_2^2}{V_2}\right) \qquad [3.4]$$

where V_1 and V_2 are the domain volumes under pressure on each side of the piston. Since these volumes vary as a function of the position of the rod relative to the body, this stiffness depends on cylinder extension. For a symmetrical cylinder of hydrostatic section area S, it is possible to show that K_{hm} reaches a minimum when the cylinder is centered which is given by:

$$K_{hm} = \frac{2B_a S^2}{V_{12}} \qquad [3.5]$$

where V_{12} is the volume under pressure on one side of the piston when it is centered.

This hydromechanical stiffness combines with the mass M of the load to create a hydromechanical natural mode of frequency ω_{hm} which can have a significant impact on dynamic performances:

$$\omega_{hm} = \sqrt{K_{km}/M} \qquad [3.6]$$

By way of example, consider a symmetrical cylinder of a single-aisle aircraft aileron. It has a hydrostatic section area $S = 4$ cm² and a total stroke of 50 mm. For a Bulk modulus B_a = 8,000 bar and considering a dead volume equal to 20% of the mean volume of a chamber, then volume V_{12} is equal to 12 cm^3. The hydromechanical stiffness of the cylinder is therefore equal to K_{hm} = 2133 daN/mm. If the aileron of inertia J = 1 kg m² is driven by a lever arm of length l = 45 mm, then its mass reflected in translation at the level of the cylinder is equal to M = 494 kg because:

$$M = J / l^2 \qquad [3.7]$$

Consequently, the frequency of the hydromechanical natural mode is ω_{hm}= 208 rad/s, in other words 33 Hz.

3.4.3. *Hydraulic inertia*

Hydraulic inertia is an energy storing/restoring phenomenon tied to the mass of matter. Inertia I_h of the fluid domain characterizes the pressure drop that is required to accelerate the fluid between two points 1 and 2 of a horizontal pipe, when the flow varies at the rate dQ/dt:

$$P_1 - P_2 = I_h \frac{dQ}{dt} \qquad [3.6]$$

This expression has the same form as the expression for electrical inductance which relates voltage to the time derivative of current. For a line of length l and of section S filled with a fluid of specific density ρ, the hydraulic inertia is given by:

$$I_h = \frac{\rho\, l}{S} \qquad [3.7]$$

Unlike hydraulic capacity, hydraulic inertia does not fluctuate much. In order to illustrate the effect of hydraulic inertia, consider abruptly steering an aileron. This creates a sudden flow rate demand of ΔQ = 20 l/min in Δt = 50 ms on a pipe of length l = 20 m and of diameter d = 12 mm with fluid of specific density ρ = 1,000 kg/m^3. In these conditions, hydraulic inertia causes a dynamic pressure drop of P_1-P_2 = 11.8 bar on the line.

3.4.4. *Speed of sound in the hydraulic fluid*

The apparent Bulk modulus B_a and the specific density ρ of the hydraulic fluid determine the speed of sound c in that fluid:

$$c = \sqrt{B_a / \rho} \qquad [3.8]$$

It becomes clear with this equation that it is possible to identify the Bulk modulus B_a by simply measuring the speed of sound in the fluid. Using common values $B_a = 8{,}000$ bar and $\rho = 1{,}000$ kg/m^3, the speed of sound in the hydraulic fluid equals $c = 894$ m/s.

4

Hydromechanical Power Transformation

4.1. Hydromechanical power transformation

Figure 4.1 provides an overview of the main families of hydromechanical transformers. They are described as motors or pumps when their range of motion is unlimited. However, when their range of motion is limited, they are described as cylinders, even for rotary motion. If functional transformation of power occurs from the mechanical domain to the hydraulic domain, then the transformer functions as a hydraulic pump. For the reverse transformation, it functions as a hydraulic motor.

Rotary cylinders with no mechanical motion conversion are hardly ever used because it is difficult to implement dynamic sealing between their moving parts. Linear cylinders are therefore most often preferred for motion conversion functions (see section 7.5.2). Moreover, they are very widely implemented because they are reliable and easy to build in many different forms:

– Single-effect cylinders can only apply a hydraulic force on the load in one direction. Motion in the other direction is provided by the action of the load on the moving part of the actuator. These cylinders are only equipped with one hydraulic port. The brakes of aircraft widely rely on the principle of the single-effect cylinder with a plunger piston in which the cylinder rod and the piston are merged together.

– Double-effect cylinders make it possible to apply hydraulic force in both directions. They are therefore equipped with two hydraulic ports. The single rod solution minimizes overall dimensions and it is well suited for actuating asymmetrical loads significantly. Therefore, from the standpoint of overall dimensions, it is naturally suitable for the actuation of landing gear (extension/retraction, doors) and spoilers. In symmetrical double rod cylinders, ring hydrostatic section areas are identical for both chambers 1 and 2 (i.e. $S_1 = S_2$). The two rods have the same diameter. This solution is well suited to quite symmetrical loads which need to be actuated in all four power quadrants (two velocity directions and two force directions). It is the method commonly adopted for steering landing gear and for primary flight controls[1]. In differential double-effect cylinders, the hydrostatic section area of chamber 1 is twice the section area of chamber 2 (i.e. $S_1 = 2S_2$). This configuration gives the ability of applying symmetrical forces while simplifying the power metering function which is only required for chamber 1, since chamber 2 is permanently connected to the pressure source. However, differential cylinders are now rarely ever used because of their low hydraulic stiffness and low effort density.

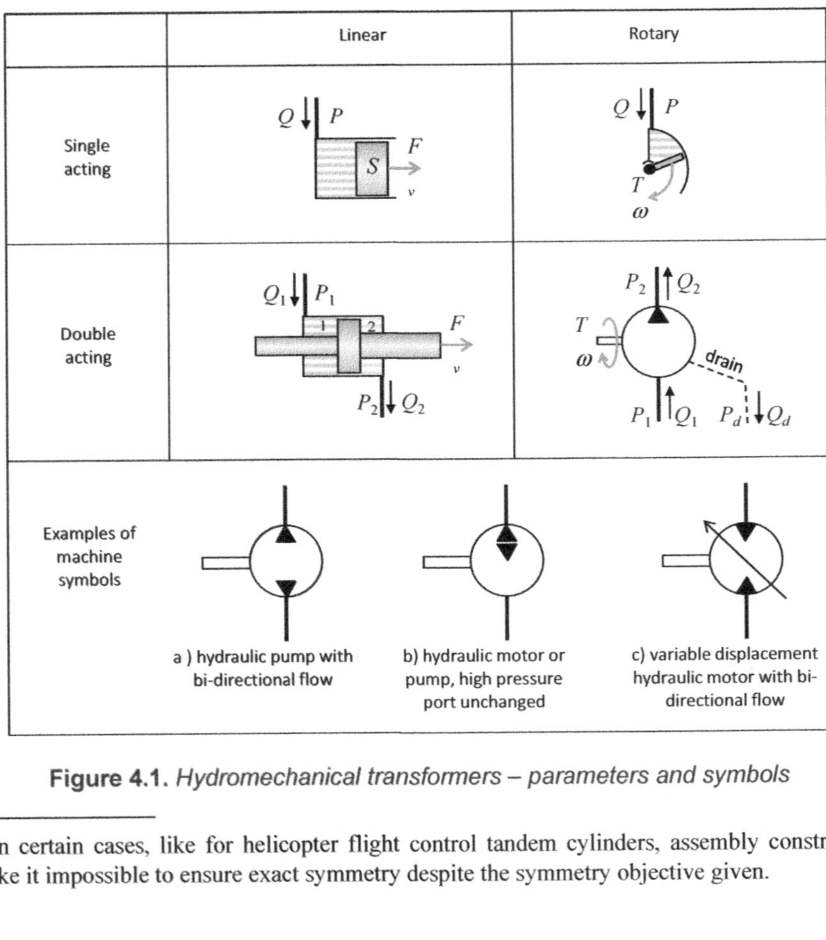

Figure 4.1. *Hydromechanical transformers – parameters and symbols*

---

1 In certain cases, like for helicopter flight control tandem cylinders, assembly constraints make it impossible to ensure exact symmetry despite the symmetry objective given.

Hydromechanical power transformation with unlimited stroke can be performed based on two major concepts:

– in *roto-dynamic* machines, the fluid is accelerated to increase its kinetic energy, which is then converted at the output into pressure energy or vice versa. This transformation is carried out under continuous matter flow, a fluid particle never being isolated from the input and output hydraulic ports of the machine;

– in *positive-displacement* (or *volumetric*) machines, the fluid is entrapped in cavities, certain walls of which are moveable. Unlike roto-dynamic machines, the flow of matter at input and output hydraulic ports is not continuous and the kinetic energy of the fluid remains globally weak during the transfer cycle.

Positive-displacement hydromechanical transformers, including cylinders, have excellent power density: up to 10 kW/kg for variable displacement pumps and even up to 15 kW/kg for certain hydraulic motors. Regarding reliability, the mean time between failures (MTBF) of variable displacement pumps ranges between 5,000 and 7,000 flight hours for military applications and between 15,000 and 30,000 flight hours for commercial aerospace.

Figure 4.2 highlights, in the form of a pressure/volume diagram, the evolution of fluid in the cavity of a positive-displacement machine operating in pump mode. The theoretical cycle is indicated by the blue dotted line. It is circled counterclockwise (abcd). Volumes V_1 and V_2 correspond to the minimal and maximal geometric volumes of the cavity defined by the stroke x which is imposed on the displacement of the piston relative to the lining. On segment ab, the cavity communicates with the domain at pressure P_1 while its volume increases: this is the suction phase. On segment cd, the cavity communicates with the domain at pressure P_2 while its volume decreases: this is the delivery phase. Work transmitted by the piston to the hydraulic cavity fluid is equal to the surface area of the abcd cycle. In practice, two major phenomena alter this transmission. They become apparent on the effective transfer cycle which is indicated by a red solid line on the diagram:

– On the one hand, part of the displaced fluid is not transferred because it is actually used to compensate fluid compressibility (meaning its volume variation as a consequence of pressure variation) during the phases of compression bc and decompression da. This effect becomes apparent in domains ❶.

– On the other hand, during the filling (or suction) phase which corresponds to segment ab, pressure P'_1 in the cavity is lower than pressure P_1 at the low pressure port. This is due to the hydraulic resistance at the input of the connection between the port and the cavity. For the same reason, pressure P'_2 in the cavity is greater than pressure P_2 at the high pressure port in the cavity during the transfer (or delivery phase) which corresponds to segment cd. The hatched domain ❷ corresponds to work lost because of these pressure drops at input and delivery.

The figure does not show the impact of leakage, which also reduces the volume transferred. It also omits showing the transient flow rates that are generated by switching (see section 4.3.4).

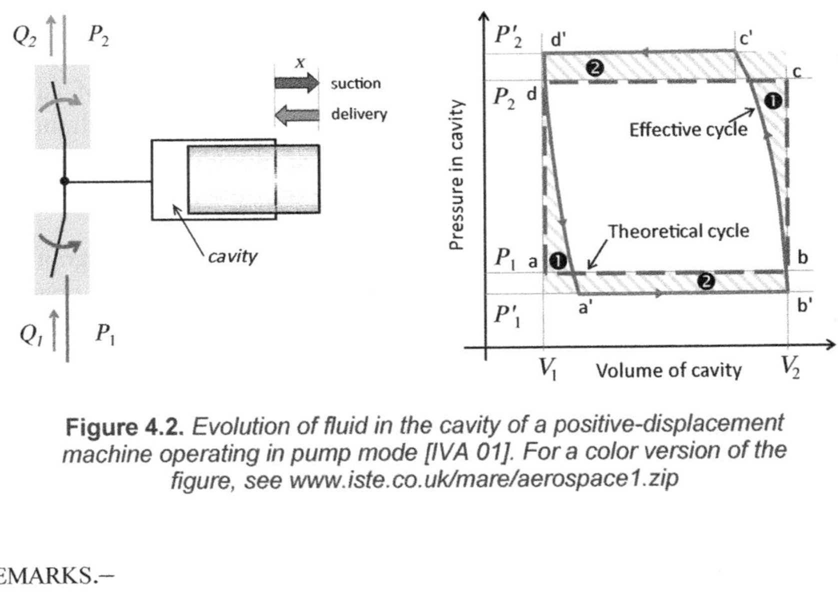

Figure 4.2. *Evolution of fluid in the cavity of a positive-displacement machine operating in pump mode [IVA 01]. For a color version of the figure, see www.iste.co.uk/mare/aerospace1.zip*

REMARKS.–

– Every cycle, the volume delivered at pressure P_2, i.e. at the use port, corresponds to the difference in volume between points c' and d'. The average flow rate Q_2 of the machine supplied over one cycle is equal to the ratio between this volume and the cycle duration. Similarly, the volume suctioned at pressure P_1 corresponds to the difference in volume between points b' and a', the average suction flow rate Q_1 being equal to the ratio between this volume and the cycle duration.

– In the case of a hydraulic motor, the cycle is circled in reverse. Therefore, pressure P'_1 is greater than pressure P_1 and pressure P'_2 is lower than pressure P_2.

Positive-displacement machines are well suited to power hydraulics which is based on energy transport in hydrostatic form. This is because: (1) the mechanical power end user usually demands high levels of effort at low speeds. However, positive-displacement machines precisely operate under moderate flow rates (from a few dozens to a few hundred l/min) and at high pressures (a few hundred bars); (2) also, as already indicated in the first chapter, because the transformation of kinetic energy into pressure energy is an inefficient process. This fact rules out roto-dynamic pumps for high pressures. However, roto-dynamic machines are not very sensitive to fluid pollution because they are equipped with very few dynamic seals. They also have a good self-suction ability. Therefore, they are most useful when these qualities are essential and when low levels of power are at play. This is why they are often used for charging (re-feeding) positive-displacement pumps.

In the next sections, particular focus will be placed on cylinders and positive-displacement machines, an example of which is shown in Figure 4.3.

Variable displacement, axial piston pump [HAL 98]

Fixed displacement, axial piston motor [CCP 05]

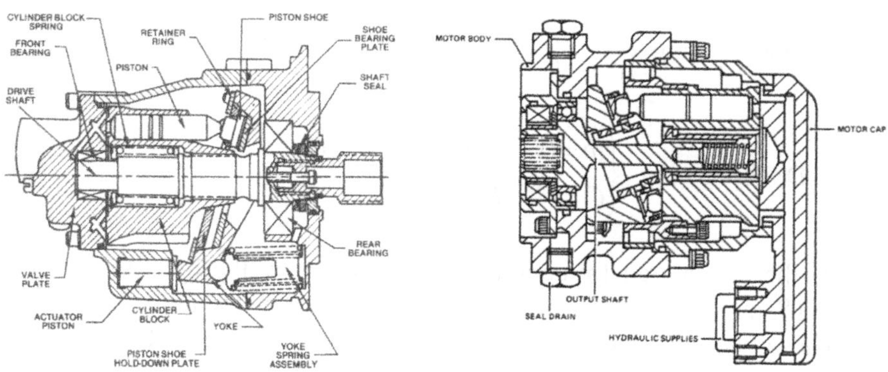

Figure 4.3. *Examples of positive-displacement hydromechanical transformers [CCP 05]*

4.2. Functional perspective

From a purely functional point of view, power transformation between the mechanical and hydraulic domains can be carried out without loss. As shown by the last line of Figure 4.1, the symbol for hydraulic machines can be upgraded to explicitly indicate for which operation quadrants the machine is designed. For this purpose, one or more specifically located triangles are added to the symbol. The triangle points outside the circle denote that hydraulic power is generated from mechanical power, and points inside denote the reverse. It also is oriented in the allowed flow direction of the fluid.

Using the notations introduced by Figure 4.1, perfect power transformation for a driving translational motion is given by:

$$\text{Single-acting} \begin{cases} v = (1/S)\, Q \\ F = S \quad P \end{cases}$$

$$\text{Double-acting} \begin{cases} v = (1/S_1)\, Q_1 = (1/S_2) Q_2 \\ F = S_1 P_1 - S_2 P_2 \end{cases} \qquad [4.1]$$

and for a rotational motion[2] it is given by:

$$\text{Single-acting} \begin{cases} \omega = (1/V_0)\, Q \\ C = V_0 \quad P \end{cases}$$

$$\text{Double-acting} \begin{cases} \omega = (1/V_0)\, Q_1 = (1/V_0)\, Q_2 \\ C = V_0 (P_1 - P_2) \end{cases} \qquad [4.2]$$

where the power transformation ratio is defined by the hydrostatic section area S (S_1 and S_2 in asymmetrical double acting) or by the unity displacement V_0. These two quantities physically represent the volume moved (in m^3) per unit of relative displacement (in m or rad, respectively) between the moving part and the machine body. For cylinders, the stroke, in other words the

2 The sign of the torque depends on whether the machine is operating in pump or motor mode. Here, the torque expressed is negative when operating in pump mode.

range of motion, is also a characteristic parameter of the power transformation function.

By multiplying the two equations that characterize each transformation together, it is possible to verify that power is functionally conserved between the mechanical and hydraulic domains. These equations, non-oriented and invertible, seem to suggest that positive-displacement machines can operate interchangeably in pump or motor mode, i.e. whatever the sign of power variables. However, technological solutions used to design these machines can be incompatible with their use in certain power quadrants. This consequently dictates specific architectural constraints.

Concerning power metering functions, it is interesting to design rotary positive-displacement machines whose displacement V_0 can be dynamically adjusted. Those are called *variable displacement* machines. However, this does not concern linear cylinders for which it is unfortunately impossible to design a variable section S. As shown on the last line of Figure 4.1, a variable cylinder is symbolized by adding an arrow on the hydromechanical power transformer symbol. This arrow can be connected to the displacement control device.

REMARKS.–

– Displacement V_0 is often (and implicitly) expressed in cm^3 or cubic inch per revolution and not per radian. Therefore, in order to perform calculations in SI units, it is imperative to make sure to convert it to m^3/rad first.

– It is often implicitly assumed that the body of the hydromechanical transformer is static. If this is not the case, the velocities v and ω involved in the previous equations are expressed with respect to those of the hydromechanical transformer body. Hence, double-acting machines are equipped in total with two mechanical power ports (rod and cylinder for translational motion, rotor and body for rotational motion) and two hydraulic power ports, and eventually the drain port (see next section).

4.3. Technological shortcomings

This section deals with the major technological shortcomings that have an impact on hydromechanical power transformation in practice.

4.3.1. *Energy losses*

Mechanical joints (prism pairs, cylindrical pairs, spherical pairs, etc.) and dynamic sealing (seals, calibrated play, etc.) inevitably lead to losses via friction and leakage. Keeping these losses under control is a serious issue for designers because there is a trade-off between friction and leakage: reducing leakage generally leads to an increase in friction and *vice versa*. Moreover, as a result of fluid viscosity variation, friction increases at low temperatures whereas leakage increases at high temperature.

In order to account for these energy losses in hydromechanical power transformers, the notions of mechanical efficiency η_m and volumetric efficiency η_v are introduced. Mechanical efficiency reflects the significance of friction and pressure drops which alter the pressure/force transformation. Volumetric efficiency reflects the presence of leakage which alters the flow rate/velocity transformation. For a double-effect symmetrical transformer operating in motor mode, if the differential pressure through the transformer is noted ΔP, the following expressions are obtained:

$$\begin{cases} v = \eta_v (1/S)\, Q \\ F = \eta_m S\, \Delta P \end{cases} \quad \text{or} \quad \begin{cases} \omega = \eta_v (1/V_0)\, Q \\ T = \eta_m V_0 \Delta P \end{cases} \qquad [4.3]$$

and when producing hydraulic power from mechanical power:

$$\begin{cases} Q = \eta_v S\, v \\ \Delta P = \eta_m (1/S) F \end{cases} \quad \text{or} \quad \begin{cases} Q = \eta_v V_0\, \omega \\ \Delta P = \eta_m (1/V_0)\, T \end{cases} \qquad [4.4]$$

In practice, efficiencies should be used carefully for two main reasons:

– Efficiencies are far from constant because they are very dependent on operating conditions (pressure, speed and temperature). Even simply relying on efficiencies at rated power is very risky for sizing the component and the hydraulic system it contributes to. This risk is equally significant when relying on thermal equilibrium which heavily depends on hydromechanical transformation energy losses.

– The use of efficiency does not allow us to account for energy losses when output power is equal to zero. This is a strict restriction, in particular in the event of Coulomb friction originating from sealing or pre-loading of mechanical transmissions and joint.

The overall effective efficiency of hydraulic pumps or motors is closely tied to the machine's technology. Its order of magnitude for axial piston machines is about 90% at the rated power for variable displacement pumps [HAL 98]. However, it quickly decreases under partial pressures, particularly for variable displacement machines: at rated speeds, the overall efficiency drops to 70% for 20% of the rated flow rate and borders on 50–10% of the rated flow rate. For bent-axis fixed displacement motors, mechanical and volumetric efficiencies can each reach 90–95% depending on the displacement.

4.3.2. *Compressibility of the hydraulic fluid*

Even though it is functionally incompressible, the actual hydraulic fluid can in fact suffer deformations (see Chapter 3). Therefore, even in the absence of leakage, the input and output volume flow rates of a positive-displacement machine can vary by a few percent, simply because of the compressibility of the fluid (the relative volume variation is typically of 1 for a pressure difference of 100 bars between the two hydraulic ports of the machine). This effect can be taken into account by introducing the concept of hydraulic efficiency η_h.

4.3.3. *Wall deformation*

Walls of fluid domains become deformed under the effect of pressure because of the compliance of the materials they are made of. This has the same effect on flow rates as fluid compressibility does: in motor mode for example, part of the exchanged flow is not used for the flow rate/velocity conversion but is actually spent on the volume variation caused by wall deformation. Moreover, the deformable walls, excited by pressure pulsations, act like a source of acoustic pollution.

4.3.4. *Pulsations*

In order to build positive-displacement machines that can operate under permanent flow or speed, or in more than one power quadrant, cavities have to be alternatively connected with each of the two hydraulic ports of the machine. This enables them to be, in turn, filled up, isolated and emptied out (see Figure 4.2 left). By design, these machines are therefore given a mechanical self-switch that depends on the relative angle between the rotor and the stator, as are the rotors of DC motors with brushes. For example, on the main pump of an airplane hydraulic circuit which is running at 3,600 rpm and that consists of 11 pistons, 660 commutations take place per second. These commutations generate flow pulsations that transform into pressure pulsations. Those are clearly visible in Figure 5.4. As illustrated by Figure 4.4, these pulsations have two distinct origins. The first is known as displacement ripple. It occurs at twice the frequency of commutations and results from the combination of active cavities. This effect can be calculated analytically quite simply for piston machines since it results from the sum of n half-sinusoids (the delivery flow rates of the n pistons) with a $2\pi/n$-phase shift. From this result, it is easy to prove that using an odd number n of cavities greatly reduces this displacement ripple. Its relative amplitude is then estimated as $125/n^2$ [TCH 79], in other words it equals 1% for 11 pistons. The second origin of the pulsations is known as commutation ripple. It appears at the same frequency as commutations and results from shocks from the sudden connection of the cavity domain to the high pressure port. Since pressures are never identical when the connection is made, a strong flow develops for a very short period of time, very often in the opposite direction to the average flow (backflow). This generates a pressure pulse in the circuit as a function of its hydraulic impedance. This effect typically causes pulsations with peak-to-peak amplitudes ranging from 10 to 20 bars [SCHM 98]. Unfortunately, the amplitude of this effect cannot be calculated as easily because it follows from transient phenomena that are difficult to model and simulate. The origins of pump pulsations have been well known for decades [KEL 74]. Therefore, several set-ups [JOH 05] aiming to minimize commutation pulsations exist (relief groove, pre-compression orifice, etc.). However, they are unfortunately only relevant in a small operation domain and they, in fact, cannot be efficiently implemented on machines with bi-directional flow. Regarding pulsation damping, the latest generation of constant pressure generation pumps are now equipped with an integral attenuator on their high pressure output, as can be seen in Figure 4.5.

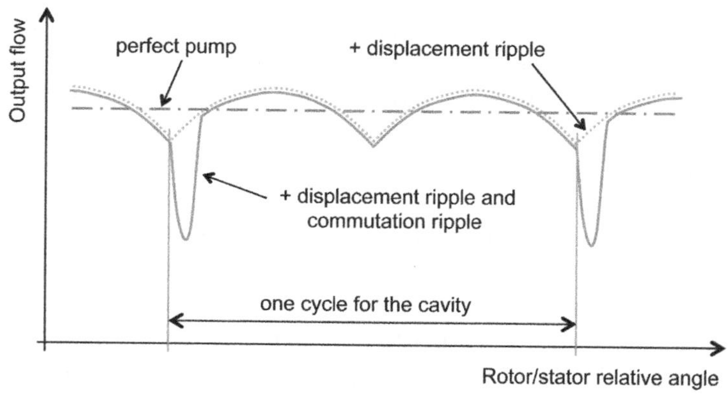

Figure 4.4. *Commutation effect*

4.3.5. *Drainage*

Most hydraulic pumps and motors are equipped with a drain orifice which has to be kept at a low pressure. Indeed, the maximum permissible pressure in the machine body should not exceed a few bars. This limits the weight and cost increase that follow from the body mechanical resistance requirement. In hydraulic pumps and motors, it is not usually possible to use seals to ensure dynamic sealing. This is due to the facts that high and permanent relative velocities quickly cause wear and that mechanical losses by friction become significant. Dynamic sealing is therefore performed by calibrated clearance whilst accepting the leakage that comes as a consequence. Since this leakage develops from high pressure areas toward low pressure areas, it tends to increase pressure in the machine body. Therefore, this body has to be drained in order to avoid a pressure build-up. In some cases, it is possible to bypass the need for an external drain orifice by internally connecting the machine body to the low pressure orifice (internal drain): the machine thus suctions leakages back in. With this solution, reversal of the flow direction is prevented by a check valve. If it is possible for the pressure difference through the machine to get reversed depending on the operation mode (hydraulic pump or motor), then the internal drain has to be connected to each of the orifices using two check valves. Nevertheless, energy losses in the machine are a source of heat that needs to be evacuated to maintain the fluid and machine temperatures within an acceptable range. In certain applications operating under permanent flow rate, the convective heat exchange between the machine and its surroundings, combined with the heat transport carried out by the hydraulic fluid leaving the machine are enough to ensure thermal equilibrium. If the machine suctions leaked fluid back in using internal drainage, thermal equilibrium is no longer ensured at high pressures (significant losses by friction) and at low flow rates (poor ability to evacuate heat by the fluid leaving the machine). Fluid flow must then be forced in the pump body to carry heat away from the pump.

For all these reasons, it is often necessary to include an external drainage function to evacuate all at once leakage flow rate, heat and solid pollution that are generated within the machine due to energy losses. An external drain orifice must then be added in order to be able to connect the machine to a low pressure domain, the reservoir for example, via an additional hydraulic line. In most cases, this additional line is fitted with a filter and a heat exchanger. The dilemma then consists of managing to rid the fluid of its pollution and cool it whilst limiting the drain pressure increase caused by pressure drops at the filter and at the heat exchanger. On pressure generation pumps, the drain flow rate typically represents between 5% and 8% of the pump rated flow rate. Furthermore, these pumps are generally equipped with a low pressure pump, of the roto-dynamic type in most cases, to enforce a minimal drain flow rate. Moreover, in the case of applications operating at 350 bars, such as on the main pumps of Airbus A380, a positive-displacement pump dedicated to force the drain flow toward the reservoir is also added [AIG 01, EAT 04].

4.4. Pump driving

On aircraft, hydraulic power generation is most often centralized. Hydraulic power is generated from pressure-compensated variable-displacement pumps. The latter can be mechanically driven in a variety of ways to satisfy the diversity requirements of hydraulic power sources.

4.4.1. *Driving performed by main engines: Engine Driven Pump (EDP)*

Some mechanical power is drawn from the main engines through the accessory transmission gearbox (see Figure 4.5). The latter is in charge of performing speed reduction as well as gear torque sharing. This is the solution used in normal mode to drive the main pumps when engines are running. Given that the rotational speed of the pump is imposed by the engine speed, it can typically range as much as fivefold depending on the phase the flight is in: for example from a speed of 800 rpm at idle to 4,000 rpm at take-off. This dependence has a significant impact on the overall dimensions of the pump with regard to its capacity to deliver a given flow rate. This is especially true in the descending and approaching phase when the engine is operating at low capacity. In this phase of the flight, high flow rates could for example be needed to make a turn, to deploy high-lift devices and extend landing gear.

However, it should be mentioned that regarding reliability, the various equipment driven by the accessory gearbox is provided with an isolation device used in the event of jamming. This ensures the independence of users. In most pumps, the drive shaft is weakened to enable it to act as a mechanical fuse in the event of over-torque. Newer pumps, such as those of Airbus A380, are equipped with an automatic de-clutching set-up that gets triggered in the event of over-torque and that can only be reset by a maintenance operation.

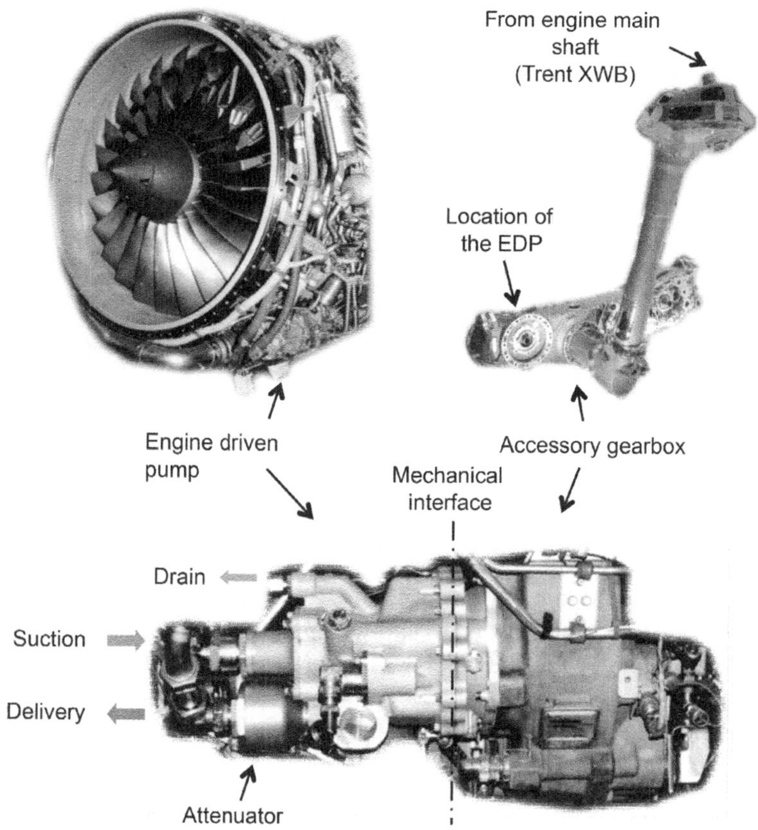

Figure 4.5. *Hydraulic pump driven by the engine via the accessory transmission gearbox*

4.4.2. *Driving performed by an electric motor: Electro Mechanical Pump (EMP) or Alternative Current Motor Pump (ACMP)*

In electromechanical pumps, the pump is driven, most often at constant speed, by an electric motor powered by the on-board electrical network (typically three-phase 115 VAC or 28 VDC), shown on the left side of Figure 4.6. In most cases, this is the solution used in backup mode. For example, in the event that one engine fails, electrical power is then supplied by one of the other motors. This solution also makes it possible to generate hydraulic power on the ground when engines are switched off and the aircraft is electrically supplied by a ground power unit.

Electro-mechanical pump [SCHW 98]
Variable displacement, pressure compensated
4.3 cm³/rev, 207 bar
7600 rpm max, 14.5 kg dry
Application: Airbus A320, A330 and A340

Local hydraulic generation [DEL 04]
Fixed displacement, speed controlled
1 cm³/rev, 350 bar
15000 rpm max, reservoir 9 l
Application: Airbus A380

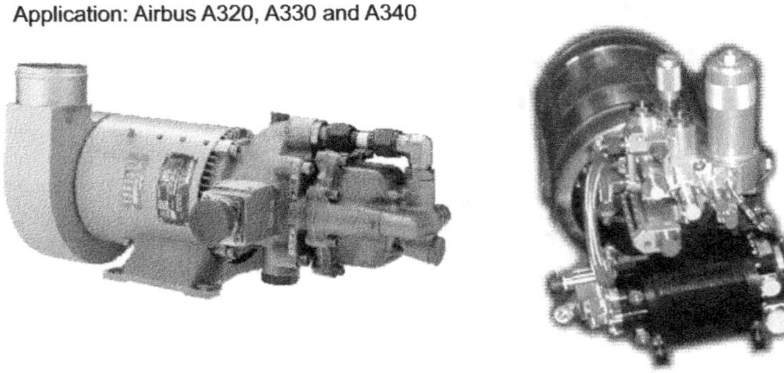

Figure 4.6. *Hydraulic pump driven by an electric motor (EDP and local hydraulic generation)*

On Airbus A380 and A350, three local hydraulic generations are used as backup for brakes and landing gear steering. Each involves a fixed displacement pump with variable driving speed as shown on the right side of Figure 4.6.

4.4.3. *Driving performed by a hydraulic motor: Power Transfer Unit (PTU)*

In certain twin-engine aircrafts, such as Airbus A320 or Boeing B737, there is a solution to satisfy reliability requirements while minimizing the on-board weight. It consists of implementing a reversible hydrostatic unit called the Power Transfer Unit (PTU). This equipment makes it possible to exchange power between two hydraulic circuits whilst complying with the segregation constraints: hydraulic power is exchanged without any fluid transfer in order to avoid the spread of leakage or pollution from one circuit to the other (see Figure 4.7). The PTU combines two positive-displacement machines, each belonging to a different hydraulic circuit. In certain applications, such as on Airbus A320, the machines can operate in pump or motor mode as needed. One of the two machines has variable displacement and the shafts of the two machines are linked. If the pressure difference between the two hydraulic networks exceeds a given threshold, the machine belonging to the higher pressure circuit operates in hydraulic motor mode. It generates mechanical power on its shaft to drive the other machine which, then, operates as a pump to power the lower pressure circuit. The power transferred is controlled by adjusting displacement, as explained in Chapter 5. This solution can be used to reduce the size of the main pumps by accepting a power input from another hydraulic network during the very short periods of high flow demands. When the PTU must be able to function with both a positive and a negative displacement, it is especially hard to reduce commutation pulsations (see section 4.3.4). As a consequence, the PTU, which is located in the belly fairing of the airplane, is clearly audible when it is activated i.e. when landing gear is retracting or when engines are switching on or off (the pumps, which are running slowly, cannot supply a great enough flow rate to fight off internal leakage of their hydraulic circuit).

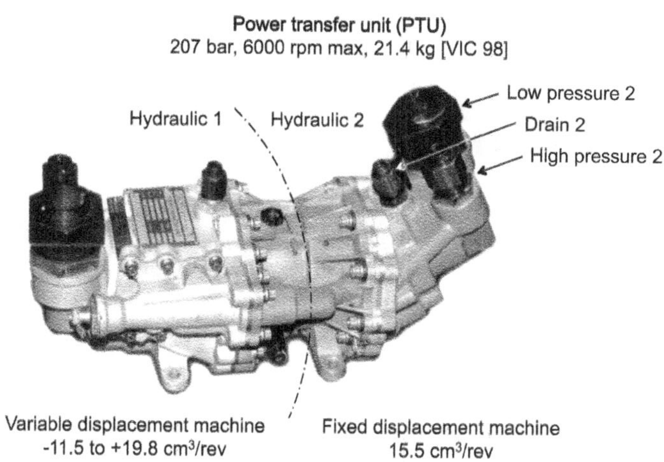

Figure 4.7. *Variable displacement power transfer unit (PTU) [VIC 98]*

4.4.4. *Dynamic air driving: Ram Air Turbine (RAT) or Air Driven Pump (ADP)*

In order to maintain crucial functions, such as flight controls in the event of a general engine shutdown (for example in the event that fuel runs out [ACC 01]), the air in motion relative to the aircraft is used to drive a hydraulic pump. However, due to its low efficiency, this solution is only used as a last resort backup, when it is no longer possible to draw power from the engines. The most common concept nowadays, consists of unfurling an autonomous ram air turbine (RAT) that has a self-adjusting pitch and whose shaft is linked to the pump shaft (see Figure 4.8 left). An alternative solution consists of using a turbo-pump air drive unit (ADP) which is powered by drawing in air from the outside. The air thereby drawn sets a turbine in rotation, whose shaft drives the hydraulic pump. All these solutions can only work if the relative speed of the aircraft compared to the air is great enough.

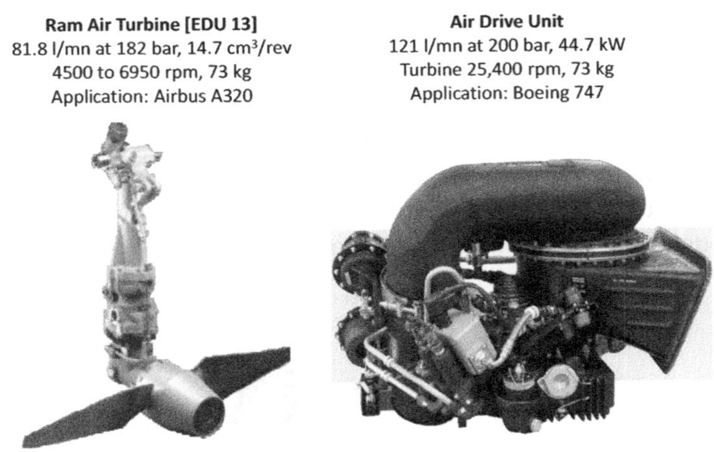

Figure 4.8. *Hydraulic pump driven by dynamic air (image on the right used with the permission of Triumph Thermal Systems, Maryland Inc.) [EDU 13]*

4.4.5. *Driving performed by a gas turbine: Solid Propellant Gas Generator (SPGG) or Monofuel Emergency Power Unit (MEPU)*

This solution consists of driving the hydraulic pump by a pneumatic turbine powered by combustion gas. The fuel can either be solid (SPGG) or liquid (EPS). Thanks to its compactness, this solution has been used as the main means of generation (SPGG) on certain missiles such as the

MX Peacekeeper. It has also been used as backup generation, for instance on fighter aircrafts F14, F15 [COW 72], F16 and Tornado. Unfortunately, the implementation of hydrazine-based liquid monofuels comes with strict security constraints because of their toxicity, which is why these solutions have been slowly disappearing.

4.4.6. *Fluid supply under pressure*

For certain applications where missions are extremely short, it is possible to bypass the use of a pump by, instead, simply relying on high pressure fluid reservoirs that are pressurized by a gas. For example, this is the case of hydraulic power generation for the thrust vector control of the solid propellant atmospheric boosters of the Ariane 5 launcher. The hydraulic fluid is pressurized at 360 bars by a helium tank. Therefore, there is no reservoir for the fluid to return to. Hence, this return fluid is lead to the nozzle output to be burnt. The main challenge remains in getting the sizing of reservoirs just right: the on-board weight needs to be minimal while guaranteeing the presence of fluid in the pressurized reservoir until the end of the mission, whatever the flight conditions.

5

Power Metering in Hydraulics

5.1. Power metering principles

Table 5.1 summarizes the different ways power can be metered in hydraulic systems. It is useful to distinguish between two main principles: Power-on-Demand and metering by energy dissipation (or valve control). As appropriate, these principles apply either to the hydraulic power source or to the hydraulic power target destination, meaning at the user level. Certain principles are compatible with the energy recovery when the load is driving. However, this recovery property is hardly ever put to use in aerospace actuation because it offers quite a small energy gain benefit while it negatively impacts the reliability and weight.

<table>
<tr><th colspan="2">Principle</th><th>Operating mode</th><th>Properties</th></tr>
<tr><td rowspan="3">Power-on-Demand</td><td rowspan="2">Metering at the source</td><td>Pump driving</td><td>R</td></tr>
<tr><td>Pump displacement</td><td>R</td></tr>
<tr><td rowspan="3">Metering at destination</td><td>Hydraulic motor displacement</td><td>R</td></tr>
<tr><td rowspan="2">Valve control</td><td>Load resistance</td><td>M</td></tr>
<tr><td>Leakage resistance</td><td>M</td></tr>
</table>

R: power regeneration ability

M: implementable with a single power source feeding several users

Table 5.1. *Power metering principles*

Power-on-Demand metering consists of acting on a variable power transformation ratio in the power chain. This can be done either at the level of the pump mechanical drive or at the pump displacement or even at the hydraulic motor displacement. From a functional point of view, this solution is attractive because it only draws the amount of power demanded by the load from the source; hence, it is called Power-on-Demand. If the architecture has been designed accordingly, it is even possible to operate in the power regeneration mode, meaning energy can be recovered when the load becomes driving. Unfortunately, the efficiency of positive-displacement machines drastically deteriorates when they operate far from their rated power or at extreme temperatures. This is illustrated in Figure 5.1, which shows that the overall efficiency of the variable displacement pump very quickly deteriorates as the displacement decreases. This phenomenon is severely penalizing since, over the course of a flight, the mean displacement of a pump is only equal to a few dozen percent of the maximum displacement.

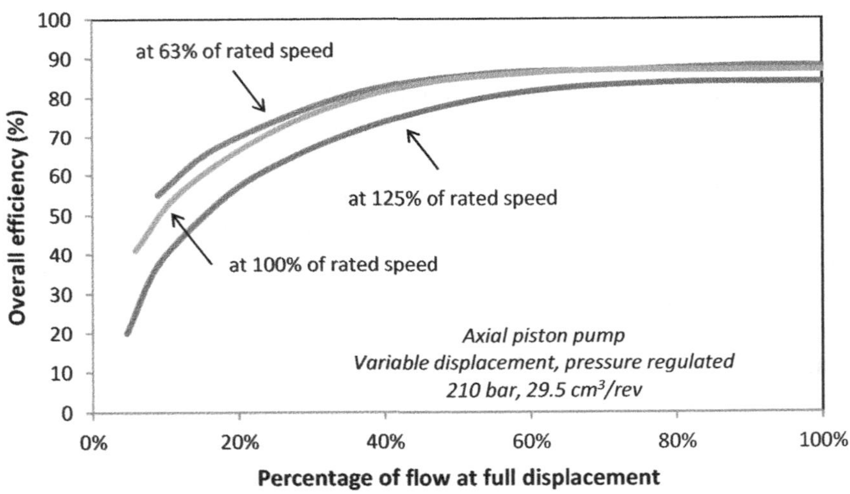

Figure 5.1. *Efficiency of a positive-displacement machine at partial load [EAT 00]*

This aspect is often poorly accounted for, essentially due to the fact that manufacturers provide little data. Indeed, their data does not cover the entire operation range (for example regarding pressure, displacement and temperature) and there is almost no data concerning the operation in regeneration mode. Furthermore, Power-on-Demand metering at the destination is not always feasible (for example, building a linear cylinder with variable cross-sectional area is not possible) and metering at the source is incompatible with the demands of several independent users, if there is only a single source.

Metering by hydraulic resistance consists of voluntarily introducing and varying leakage hydraulic resistances (in parallel) or load hydraulic resistances (in series) on the power path between the source and hydraulic loads. In principle, this solution is not very attractive: the source supplies more power than the load demands and this excess power is wasted in the form of heat. It also prohibits regeneration. However, metering by hydraulic resistance is the only solution suitable for architectures with centralized hydraulic power and with multiple users of the cylinder type. Furthermore, the steep efficiency drop that the hydraulic motors and pumps undergo at partial load or speed and at extreme temperatures greatly diminishes the energy advantage that Power-on-Demand metering has over metering by hydraulic resistance [MAR 09].

REMARKS.–

– In terms of power metering, applying the on-demand metering principles used in power electronics has also been considered. This requires setting up on/off connections between the power source and the load by means of a pulse width modulation controller. Unfortunately, in spite of all attempts, this solution has never been achieved because it is ill-suited for power hydraulics. First, it generates strong pressure pulsations that are not compatible with the permissible level of noise. This quickly causes high levels of fatigue for materials. Second, in order to build the on/off hydraulic switches, it is necessary to shift moving elements with frequencies quite high compared to the dynamics required by the load. This is incompatible with the constraints on power, wear, weight and overall dimensions.

– In a hydraulic restriction, lost hydrostatic energy is converted to heat. The maximum temperature rise of the fluid as it flows through the restriction can be calculated under the assumption that all heat, thereby generated, is transmitted to the fluid. It follows from equations [1.6] and [1.7] that this temperature rise can be expressed as:

$$\Delta\Theta = \frac{1}{\rho C_p}\Delta P \qquad [5.1]$$

This proves that the maximum temperature rise only depends on physical properties of the fluid and on the pressure drop. For common hydraulic fluids, this temperature rise is approximately equal to 6°C for a pressure drop of 100 bars.

5.2. Power-on-Demand

5.2.1. *Metering by pump drive adjustment*

This metering method consists of acting on either the rotational speed or the drive torque of the pump. This principle works when the mechanical power source driving the pump can be adjusted with sufficient accuracy and dynamic. In fact, it is ill-suited for the propulsion engine driving and for the applications where the mechanical source is not exclusively dedicated to driving the pump. However, this method makes it possible to take full advantage of the performance of modern day electrical drives. In this case, power electronics, associated with an electric motor, functionally acts as a variable power transformer between the electrical power network and the motor. The transformation ratio to be implemented is realized by flight control electronics. The goal is to servo the electric motor shaft; hence, the pump shaft, in transmitted speed or torque. For example, the speed servo-loop on the motor makes it functionally possible to act on the pump flow rate, as shown in equation [4.2]. Pressure, therefore, develops as a function of the hydraulic load to which the pump is connected. This load also determines how the pressure depends on flow (hydraulic impedance). Pressure is reflected in the mechanical domain by a motor torque demand. Therefore, from a causal standpoint, the following expressions can be obtained for this example:

– flow rate Q generated by the pump with displacement V_0, as a response to the rotational speed demand ω:

$$Q \Leftarrow V_0\ \omega \qquad [5.2]$$

– torque T reflected at the driving motor by the pump which generates a differential pressure ΔP:

$$T \Leftarrow V_0\ \Delta P \qquad [5.3]$$

The control unit of the electric motor relies on the power electronic components (IGBT, MOS-FET, etc.) that run at high frequency commutation (a few kHz). The motors also include high performance magnets. These give the electric drive excellent overall efficiency at rated power. Concerning power electronics, energy losses mainly result from conduction resistance and parasitic capacity when switching. Regarding the motor, they originate from the winding resistances, Eddy currents, magnetic saturation and magnetic hysteresis. As for the dynamic response of power metering, it is most often limited by the inertia of rotating parts of the electric motor and pump.

After more than 10 years being researched and matured, this solution has finally been introduced on Airbus A380 [MAR 04]. It has been applied to the electro-hydrostatic actuators (see Table 1.4) and has made it possible to remove one of the centralized hydraulic circuits.

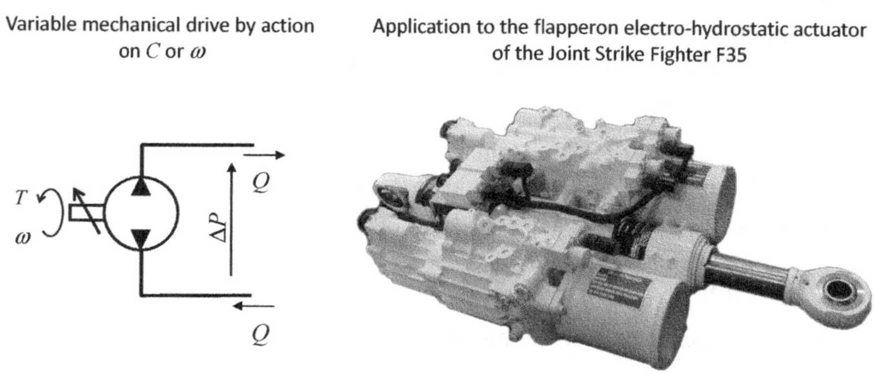

Figure 5.2. *Power metering by action on the pump mechanical drive*

5.2.2. *Metering by displacement adjustment*

An alternate version of Power-on-Demand metering consists of acting on the displacement of a positive-displacement machine. Metering by displacement adjustment can be carried out at the source, where hydraulic power is generated by soliciting a variable displacement pump. It can also be carried out at destination, in other words, where the hydraulic power is transformed into mechanical power, if a variable displacement hydraulic motor is used.

In practice, in order to modify the displacement of a positive-displacement machine, a moving organ of this machine has to be actuated. Even though this function requires very little power compared to the power of the machine, it should not be neglected while looking at power architectures. Moreover, a variable displacement machine is always more complex than a fixed displacement machine. It, therefore, has lower reliability than the latter.

The fact that the parts to be moved in order to adjust displacement have a small inertia encourages a low response time, as shown in Figure 5.4. In practice, the inherent dynamics of the displacement adjustment actuator heavily impact the overall dynamics of the power metering function.

Displacement adjustment at the source has been used for several decades in pressure-compensated pumps. Displacement is imposed by a hydraulic piston controlled by a hydromechanical regulator. This set-up aims to maintain the pump output pressure constant [FAY 91]. The response time of displacement adjustment is of a few tenths of a second. For this metering principle, the consequent dependencies can be functionally expressed by the three following relationships:

– action on pump displacement V_0 as a function of the error between the pressure difference to be generated ΔP^* and the pressure difference ΔP generated by the pump:

$$V_0 \Leftarrow f(\Delta P^* - \Delta P) \quad [5.4]$$

– flow rate generated by the pump:

$$Q \Leftarrow V_0\,\omega \quad [5.5]$$

– hydraulic impedance of the load:

$$\Delta P \Leftarrow g(Q) \quad [5.6]$$

– torque demanded by the pump:

$$T \Leftarrow V_0\,\Delta P \quad [5.7]$$

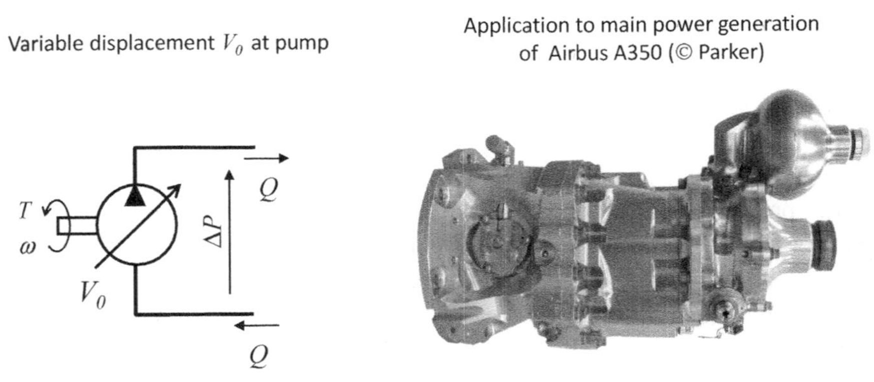

Figure 5.3. *Power metering by action on displacement at the source*

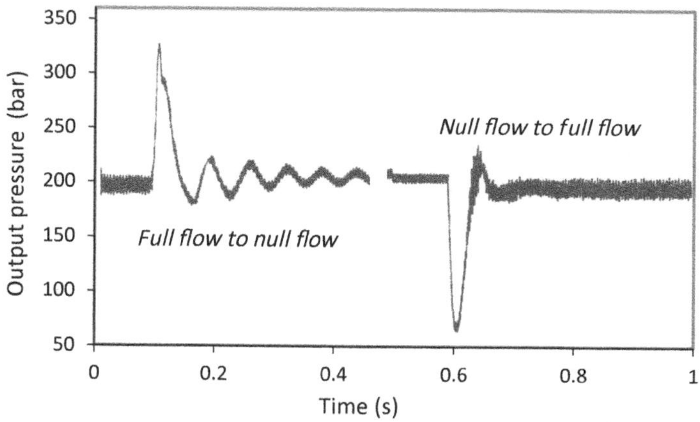

Figure 5.4. *Example of dynamic response of a variable displacement, pressure-compensated pump (main power generation of a commercial aircraft)*

Displacement adjustment at destination is used to hydraulically drive constant-frequency AC generators (Constant Speed Motor Generator, CSMG). This solution is implemented on single-aisle twin-engine airplanes, such as Airbus A320. Electrical power is generated in the event of a general engine shutdown by the CSMG. In that case, hydraulic power is supplied to the CSMG by the backup ram air turbine (see section 4.4.4). Displacement adjustment at the destination is also interesting for secondary flight control actuators because it considerably diminishes energy consumption [LIN 86, BIE 98]. It was introduced on Airbus A380 [BOW 04] for the actuation of flaps and slats. Those used to be controlled by variable hydraulic restrictions. Power is supplied at constant pressure to flaps and slats actuators and their respective position is placed in a servo-loop by action on the hydraulic motor displacement. In this example, the consequent dependencies between power variables are given functionally by the following equations:

– Action on motor displacement V_0 as a function of the error between the expected angular position θ^* and the measured position θ:

$$V_0 \Leftarrow f(\theta^* - \theta) \quad [5.8]$$

– Torque generated by the hydraulic motor on the mechanical load:

$$T \Leftarrow V_0 \, \Delta P \quad [5.9]$$

– Load mechanical impedance seen by the hydraulic motor:

$$w \Leftarrow g(T) \qquad [5.10]$$

Flow rate demanded by the hydraulic motor:

$$Q \Leftarrow V_0\,\omega \qquad [5.11]$$

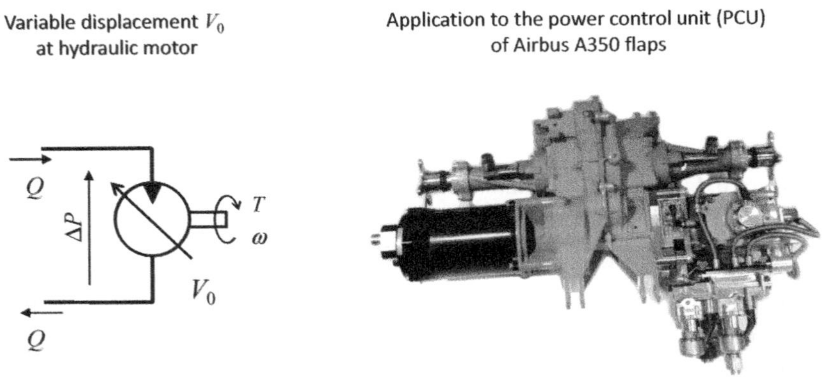

Figure 5.5. *Power metering by action on displacement at destination*

5.3. Metering by hydraulic restriction

Metering by hydraulic restriction consists of inserting one or more hydraulic resistances between the hydraulic power source and each user. Two generic solutions can be distinguished, as shown in Figure 5.6 (these solutions can also optionally be combined):

– voluntarily create leakage Q_b between high and low pressure lines of the source, by mounting the hydraulic resistance in parallel;

– create pressure loss P_d on the lines connecting the source to the user or the user to the reservoir return, by mounting the hydraulic resistance in series.

With regard to the metering function, restrictions are variable in most cases. Their control signal, *u*, will be noted here, no matter the operating mode adopted (electrical, hydraulic, mechanical or pneumatic). Hydraulic energy loss in these resistances is, therefore, functional (or structural). Unlike Power-on-Demand metering, this energy loss does not result from a technological implementation shortcoming but from its very concept. The main architectures that are derived from it can be simplified as a few generic set-ups based on considerations that are detailed below.

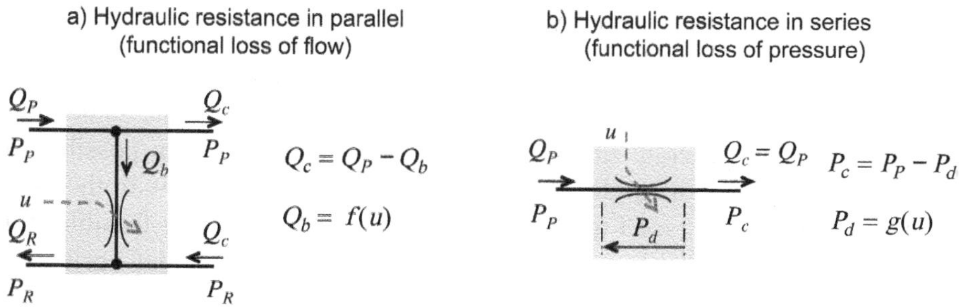

Figure 5.6. *Concepts of power metering by variable hydraulic resistance*

5.3.1. *Functional configuration*

Connections between the hydraulic power source (supply and reservoir return) and the load are set up by hydraulic valves that have two or three functional configurations (or even more in some specific cases). These are described in Figure 5.7.

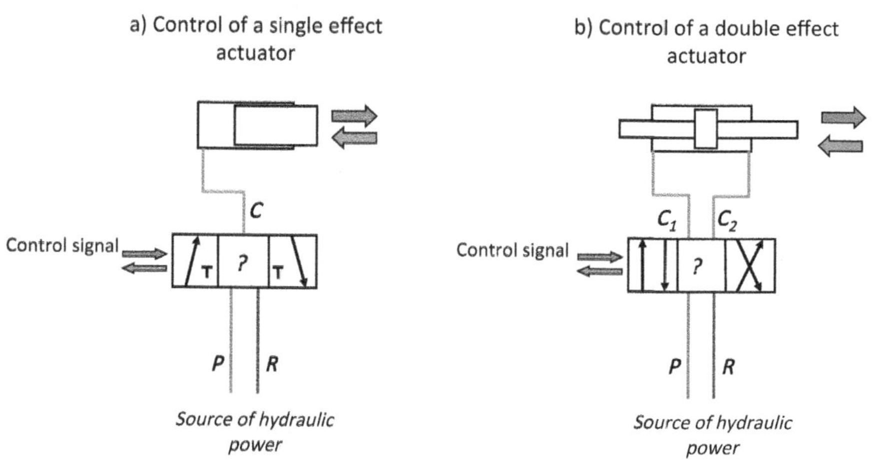

Figure 5.7. *Power metering for a single or double effect hydraulic load*

In terms of symbols, each possible configuration is represented as a rectangle in which connections (channels) between hydraulic ports (orifices) are drawn. The control signal is depicted on the other side of the rectangle. When graphic representation is possible, the active configuration is the one with the rectangle on which the control signal is applied.

5.3.1.1. *Single-effect and double-effect hydraulic user*

In certain applications, it is only required to supply power to the load in two quadrants: the hydrostatic force to develop is always of the same sign and the speed to generate can be positive or negative. It is then possible to only meter power on a single hydraulic line. This can be done with the help of a three-port valve (see Figure 5.7 left). This solution is only feasible if a driving force is available to ensure retracting (internal cylinder spring or aiding load). Depending on the opening of the valve, the control port C is connected either to the pressure source P or to the reservoir return R. This solution is, for example, used for the single-effect cylinders in brakes: their pistons are pulled back in by a spring.

When power must be metered in all four quadrants, two hydraulic lines have to be used for metering. This makes it possible to reverse both the flow (and consequently, the load speed) and the differential pressure (and consequently, the hydrostatic force applied to the load). In this case, a four-port valve is used. This allows the two ports C_1 and C_2 of the hydraulic load to be connected to the power source P and to the reservoir return R (see Figure 5.7 right). This is the most common configuration. It is encountered in flight control actuators, landing gear steering and extension-retraction, as well as engine thrust reversers.

Therefore, the hydraulic valve has an important directional role: it introduces or removes hydraulic ways that define the functional path of power.

5.3.1.2. *Intermediate state*

The intermediate state, depicted in Figure 5.7 by a question mark, does not exist in the case of a two-state valve where power can only go in two directions ($P \rightarrow C$ or $C \rightarrow R$ for a single-acting load, $P \rightarrow C_1$ / $C_2 \rightarrow R$ or $P \rightarrow C_2$ / $C_1 \rightarrow R$ for a double-effect load). However, having an intermediate configuration is most often needed to satisfy the requirements and constraints on the load or on the hydraulic power source. Figure 5.8 introduces several different common configurations.

Figure 5.8. *Different configurations of center for hydraulic valves*

From the point of view of the hydraulic load, solutions a–d make pressures at ports C_1 and C_2 equal. The hydrostatic force is, thus, cancelled if the hydraulic load is symmetrical. Then, the load can be driven by an external force, aside from effects due to seal friction and inertia of the moving part of the actuator. These four solutions differ in the coupling they create between the load and hydraulic generation. Ports C_1 and C_2 are connected to the pressure source in solution a, to the return in solution b, to both pressure and return in solution c and they are isolated in solution d. Solutions e and f isolate the load. The latter is then locked in position, aside from effects due to leakage and hydraulic fluid compressibility.

From the point of view of the hydraulic generation, ports P and R are bypassed in solutions c and e and they are isolated in the other solutions. The importance of these configurations will be described in the following paragraphs.

5.3.1.3. *Directional and metering functions*

In certain applications, the actuator operates in the on/off mode. For example, this is true for landing gear extension and thrust reverser deployment. When this is the case, no attempt is made to control the load in real time. Instead, point-to-point movements are triggered, or rather, end-stop to end-stop movements. Then, it becomes possible to dissociate the directional function from the metering function. The metering function can be implicit, as in the case shown on the left of Figure 5.8. In this case, it is the inherent hydraulic resistance of the directional function which determines the load speed as a function of its impedance. Fixed hydraulic resistance can also be inserted between the directional valve and the hydraulic load, as shown in the center of Figure 5.9. In order to safeguard against the cavitation and desorption effects when the hydraulic fluid pressure drops, flow toward the chamber that is filling up should not be restricted. This is because this chamber acts as a syringe if the load is driving. The hydraulic resistance is, therefore, placed on the line of the chamber that is emptying out. In order to satisfy this safeguard for the opposite flow direction, a check valve is generally mounted in parallel with the hydraulic restriction to bypass it. Moreover, this solution makes it possible to dissociate speed adjustments for the two load movement directions.

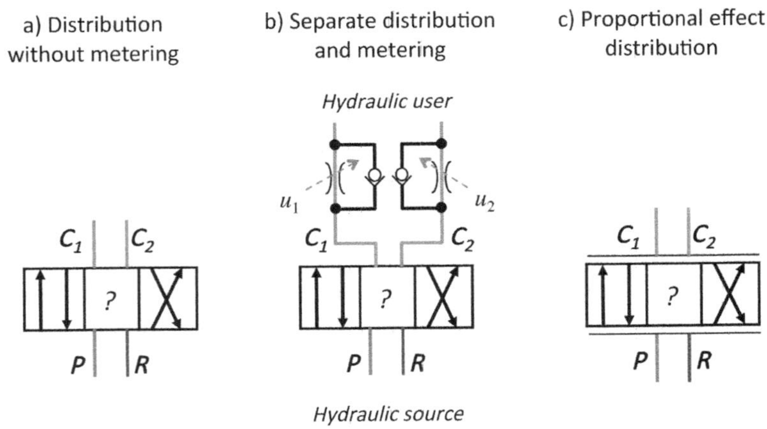

Figure 5.9. *Power distribution and metering*

5.3.1.4. *Metering carried out by a flow control valve*

When metering must be carried out without reversing the direction of the flow, it is possible to use flow control valves (or flow regulators), as described in Figure 5.10. By design, flow regulators involve a pressure regulator. Its purpose is to maintain a constant differential pressure across a hydraulic resistance mounted in series on the flow that has to be regulated (resistance numbered 1 on the diagrams). The static hydraulic characteristic curve of the restriction then defines the relationship between the flow across it and the differential pressure it creates. If the source is a permanent flow source, configuration c should be used in order to redirect the excess flow toward the return line.

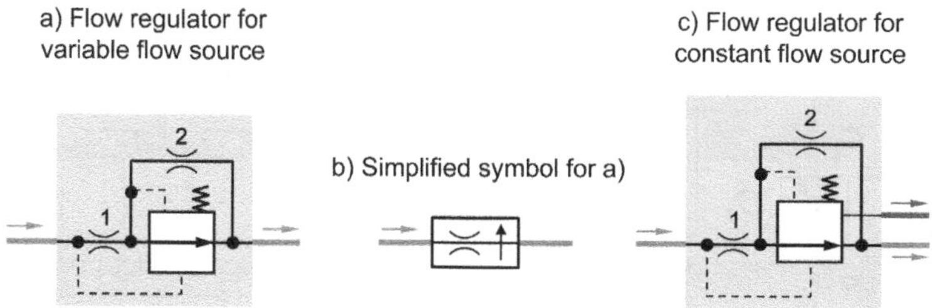

Figure 5.10. *Flow control valves*

5.3.1.5. *Metering carried out by a proportional valve*

In the majority of actuation applications, power metering is not known ahead of time and depends on the pilot's setpoint. A possible solution could consist of having the control signal modulate the restrictions of flow control valves. This solution is well-suited to meter unidirectional flow such as in fuel metering. However, it is not used for actuators. Indeed, it is ill-suited to fine-tune the load position at close to zero speed because the flow direction can fluctuate dynamically. The directional and metering functions therefore have to be coupled. This is done by implementing proportional valves in which the cross-sectional area of the fluid path can be continuously varied (see Figure 5.9 right). On hydraulic schematics, proportional valves are differentiated from directional valves by means of a double line above and below their rectangle.

Four-way proportional valves (shown on the left of Figure 5.11) may be viewed as the hydraulic equivalent to a variable four-way Wheatstone bridge (shown in the center of Figure 5.11). The architecture here is similar to the H-bridge used in power electronics, although there is one substantial difference between the two. Indeed, when paths a and b are progressively opened, paths c and d are functionally closed and vice versa. Three-way valves used for singe-effect hydraulic loads (a single controlled chamber) only form a half-bridge.

Most proportional valves are built from the combination of a cylindrical spool with a cylindrical sleeve, as can be seen on the diagram on the right of Figure 5.11. An example of the tandem version of this set-up (double in-line four-way valve) is shown in Figure 5.12 for the inboard aileron actuator of Boeing B747-8.

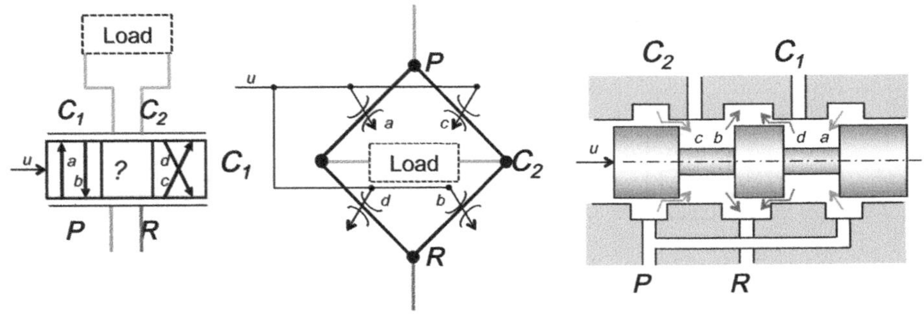

Figure 5.11. *Four-way proportional valve, viewed as a Wheatstone bridge*

Spool and sleeve of inboard aileron actuator main valve
(Boeing B747-8)

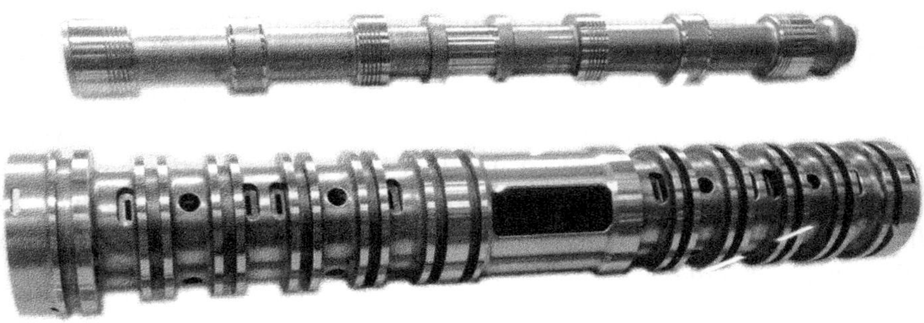

Figure 5.12. *Example of an aerospace hydraulic proportional valve design*

5.3.2. *Types of distribution*

5.3.2.1. *Series distribution*

Historically, hydraulic power sources relied on fixed displacement pumps, for the sake of simplicity. When they were driven in rotation, the pumps would continually supply flow. As a consequence, the flow in excess with respect to the users' needs had to be re-circulated to the low pressure line, as described in Figure 5.13.

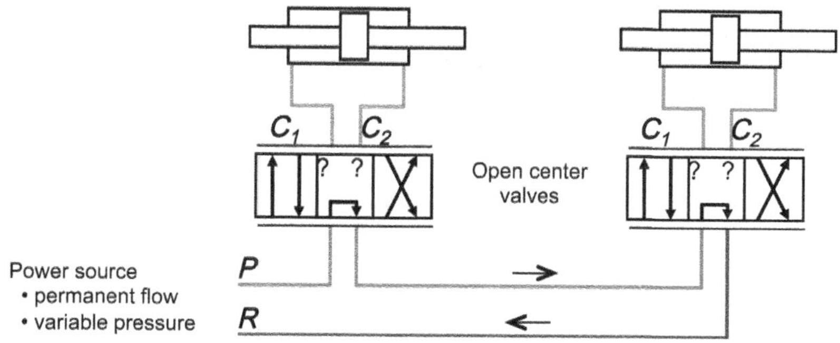

Figure 5.13. *Series distribution*

In order to set-up series distribution, an open center valve is required. Its purpose is to redirect the flow from the source to the user when appropriate instead of letting the flow return to the reservoir. Since valves are associated in series on hydraulic lines, the pump output pressure is functionally equal to the sum of the pressures demanded by users. In the case of an asymmetrical cylinder, the flow rate in the lines is proportional to the ratio of cylinder hydrostatic cross-sectional areas. In some cases, a hydraulic restriction is inserted on the return line. This is done to ensure a minimal level of pressure that can be used for secondary functions.

This solution was abandoned for two fundamental reasons. First, it introduces strong coupling between users. Secondly, it has very poor dynamics, since generating a force at the level of users requires the entire upstream supply line to be pressurized. Additionally, as previously mentioned, the conversion of kinetic energy into pressure energy, which is widely put to use here, is a very inefficient process.

5.3.2.2. *Parallel distribution*

When the hydraulic power source makes it possible to run at zero flow, closed center valves mounted in parallel can be used. This time, the hydraulic resistance acts in series on both coming and going flows of the user.

All valves see the same supply pressure. Additionally, the output flow rate of the pump is functionally equal to the sum of flow rates demanded by each user. This solution has only started being widely used since the pressure-compensated, variable displacement pump technology has been mastered[1]. It generates little interaction between users and it offers excellent response dynamics. Moreover, the conversion of the pressure energy in the hydraulic supply line into kinetic energy is performed with very good efficiency.

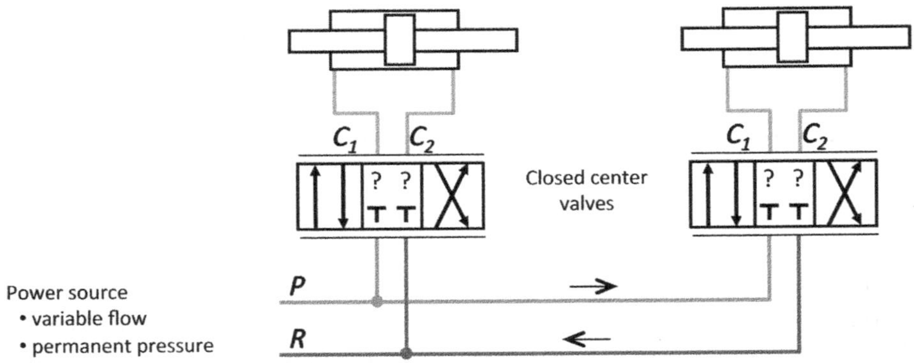

Figure 5.14. *Parallel distribution*

The aim is to build efficient position servo-loops ensuring a good rejection of force disturbances externally applied on the load. In order to achieve this, a closed center is added at the C_1 and C_2 ports (solution f on Figure 5.8). In the end, most position servo-loops nowadays rely on a closed center proportional four-way valve. When the controls are electrical, they come in the form of direct-drive servovalves (see section 5.5).

5.4. Impact of restriction configuration and properties on the metering function

5.4.1. *Fixed hydraulic restriction*

Ideally, for a hydraulic resistance, differential pressure ΔP would vary linearly with flow Q (as for an electrical resistor where the voltage across is proportional to the current flowing through). This would make the designers' and, more so, the automaticians' jobs much easier. However, in real-life hydraulics things are far from ideal, as can be seen in Figure 5.15.

1 These pumps have been used in commercial aerospace since the late 1950s, such as on the Boeing B707 for example.

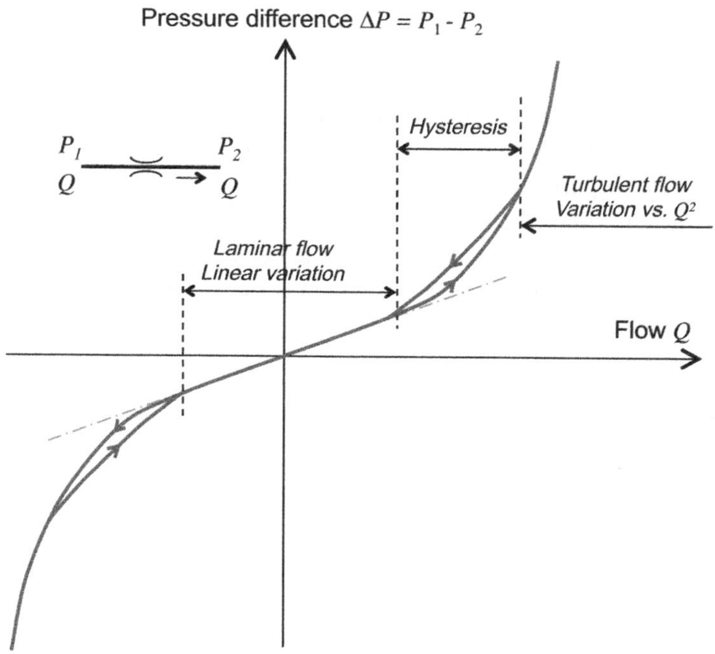

Figure 5.15. *Static hydraulic characteristic curve (flow vs. differential pressure) of a hydraulic restriction*

5.4.1.1. *Laminar flow*

Pressure varies linearly with flow only for low flows and/or for low temperatures which lead to laminar flow. Energy loss is then the result of friction between hydraulic fluid particles. Hydraulic resistance in laminar flow regime is directly proportional to the dynamic viscosity μ:

$$\Delta P = k_l \, \mu Q \qquad [5.11]$$

However, the viscosity of the hydraulic fluid is extremely sensitive to temperature: viscosity typically gets divided by 10 between −40 and 0°C and is further divided by 10 between 0°C and 100°C. This sensitivity is an issue in most cases because it alters the hydraulic resistance function. Nevertheless, it can be turned into an advantage in some cases. Moreover, factor k_l is seldom proportional to the cross-sectional area at the restriction of the fluid path, which makes it difficult to implement variable cross-sections.

5.4.1.2. *Turbulent flow*

When flow rate or temperature increases, the flow becomes turbulent. The energy loss in the hydraulic resistance then results from soft shocks between fluid particles that move in an erratic manner. Part of the initial kinetic energy of particles is dissipated as heat at the moment of impact. In turbulent flow regime, the pressure drop across the hydraulic resistance (see equation [5.13]), varies with the square of the mean flow velocity. In other words, it varies with the square of the ratio between flow Q across it and cross-sectional area s it offers the fluid:

$$\Delta P = k_t \frac{\rho}{2s^2} Q^2 sign\ (Q) \qquad [5.13]$$

This pressure drop across the hydraulic resistance is also proportional to the fluid specific density ρ. The latter is, unlike other parameters in the equation, relatively insensitive to temperature and pressure changes (typically varies by 10% on the usual working range). The reciprocal form of equation [5.13] gives the hydraulic resistance in conductance form. It then clearly appears that flow is proportional to path cross-sectional area s:

$$Q = \frac{s}{\sqrt{k_t}} \sqrt{\left|\frac{2\Delta P}{\rho}\right|} sign\ (\Delta P) \qquad [5.14]$$

In order to characterize a hydraulic restriction in turbulent flow, one of two coefficients is usually introduced: either the turbulent pressure loss coefficient $\xi = k_t$ or the flow coefficient C_q ($C_q = 1/\sqrt{\xi}$). This choice depends on whether the resistance or the conductance form is used.

5.4.1.3. *Transition*

The transition between the two flow regimes – laminar and turbulent – is paired with a hysteresis effect that is difficult to assess and control. Since the flow is heavily disrupted by the hydraulic restriction (sudden changes in direction or velocity), the turbulent flow regime develops easily. Hence, this transition occurs for flow rates 5 to 10 times lower than flow rates required to trigger the transition in a straight tube with the same cross-sectional area as the restriction. In other words, it occurs for Reynolds numbers typically ranging from 100 to 200 instead of 1,500 to 2,000.

The laminar, hysteresis and turbulent domains also appear when the flow and differential pressure change signs. If the geometry is not symmetrical, then the hydraulic characteristic curve of Figure 5.15 is not exactly symmetrical about the origin [TCH 79]. This effect is poorly documented and under-researched.

Finally, it appears that in order to limit the sensitivity of the hydraulic resistance to temperature changes, it is best to strictly use hydraulic restrictions in turbulent flow conditions. However, this entails losing the linearity of pressure with respect to flow.

5.4.1.4. *Orders of magnitude*

It is interesting to illustrate the effect of hydraulic resistance with a few examples. For this, Table 5.2 provides flow rates calculated for a phosphate-ester fluid (with properties according to standard SAE-AS1241) for different geometries creating a pressure difference of 300 bars.

Type of orifice	Dimensions	Flow rate at ΔP = 300 bars (l/min)	Dissipated power
Short cylindrical orifice	Length 1 mm Diameter 0.2 mm	0.233 at −15°C 0.332 at +70°C	117 W at −15°C 166 W at 70°C
Pump spool/sleeve ring clearance	Length 20 mm Diameter 10 mm Radial clearance 0.01 mm	0.00215 at −15 C 0.03850 at +70°C	1.1 W at −15°C 19 W at +70°C
Cylindrical valve rectangular slot	Stroke 1.22 mm four slots of width 1.4 mm	72.0 at −15°C 74.6 at +70°C	36.0 kW at −15°C 37.3 W at +70°C

Table 5.2. *Orders of magnitude for hydraulic restrictions*

5.4.2. *Variable hydraulic restriction*

5.4.2.1. *Types of restrictions*

In order to make hydraulic resistances variable, the fluid path cross-sectional area s that the restriction provides for flow, is acted upon as a function of the relative opening x (translation) or θ (rotation) of the restriction. Two major design concepts can be implemented in order to achieve this [BLA 60]. They are schematically represented in Figure 5.16[2].

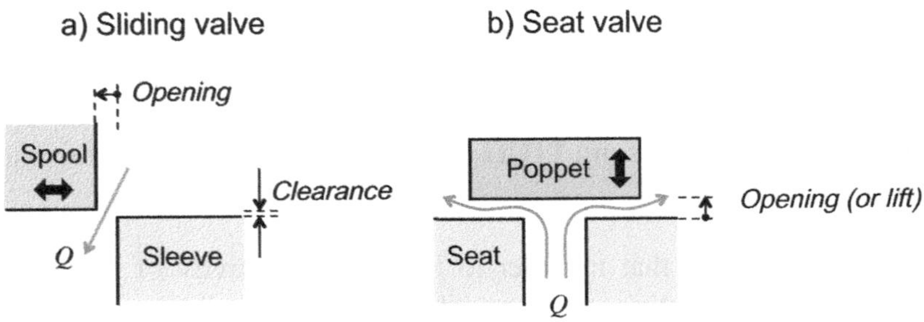

Figure 5.16. *Two concepts of variable orifice construction*

In a *sliding* restriction, depicted on the left of Figure 5.16, the fluid is subjected to shear when the relative opening decreases. This type of restriction is widely used in proportional metering for power transmission applications. Directional and metering functions are inherently combined, pressure gain at zero flow is high and hydrostatic balance is easily achieved in order to minimize driving forces. The relative opening can be negative when the restriction is functionally closed, for example when wanting to construct a proportional hydraulic valve. In this case, the existence of a clearance between moving parts is vital. However, as a consequence, perfect sealing cannot be achieved, unless seals are used, but this would be incompatible with the accuracy and the stability of the metering function.

In a *seat* restriction, depicted on the left of Figure 5.16, the fluid is pinched when the relative opening decreases. This type of restriction makes it hard to combine directional and metering functions, especially if the flow has to be able to change signs. As a result, seat restrictions are well-suited for thermo-hydraulic systems for heat transport. The poppet is difficult to balance from a hydrostatic force perspective. However, since perfect sealing is easy to achieve between the poppet and its seat, these variable restrictions are very widely used in power hydraulics for protection functions, such as in pressure relief valves. The relative opening cannot be negative considering the poppet comes into contact with its seat at zero lift.

2 There is also a third "flow sharing" type, but it has been excluded from this list because it involves two hydraulic resistances. Given it is used only as a force amplification device for valve control, it will be addressed in the sections dedicated to servovalves for the controlling stages.

In order to simplify the manufacturing process of variable restrictions, in most cases their designs are based on solids of revolution: cylindrical spool and sleeve (see Figure 5.12) for sliding restrictions and cylindrical poppet and seat for seat restrictions (vertical axis of symmetry on Figure 5.16 right).

5.4.2.2. *Geometrical characteristics of a variable restriction*

As shown by equations [5.13] and [5.14], the static characteristic curve of a hydraulic resistance ties the three variables s, ΔP and Q together.

The geometric characteristic curve is what defines the relationship between the opening of the restriction (x in translation or θ in rotation) and the cross-sectional area of the fluid path s. Functions $s = f(x)$ or $s = f(\theta)$ strongly contribute to the properties of variable restrictions with regard to the metering function. To this end, they are given specific profiles. The most generic ones are depicted in Figure 5.17.

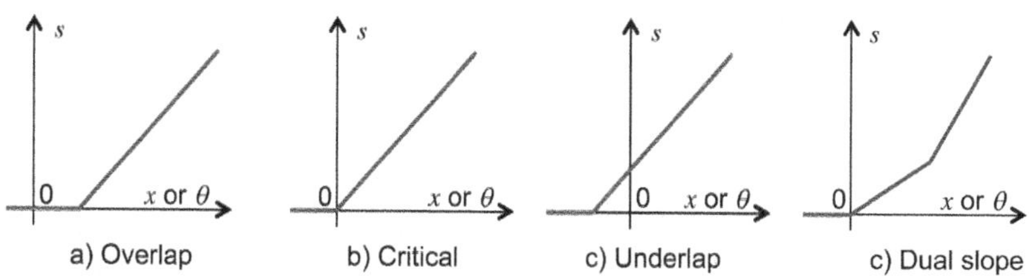

Figure 5.17. *Generic geometric characteristic curves of variable orifices*

For a *critical* restriction, the cross-sectional area of the fluid path varies strictly linearly with the relative opening. This solution is generally used in high performance position servo-loops. It is the method that generates the highest pressure gain when the cross-sectional area approaches zero. This facilitates the loaded start-up (with zero control flow, a valve opening of 1% of the rated value typically creates at the load a differential pressure of 40% the supply pressure).

For an *underlapped* restriction, the cross-sectional area of the fluid path varies linearly with the displacement of the restriction moving part. However, it is not equal to zero for zero relative displacement (the underlap), which corresponds to a negative bias. This solution reduces the pressure gain when the cross-sectional area approaches zero. It is, for instance, used for the actuation of landing gear steering in order to improve the stability of the position servo-loop.

For an *overlapped* restriction, the cross-sectional area of the fluid path also varies linearly with the displacement of the restriction moving part. However, this is only the case from a minimal relative displacement (the overlap), which corresponds to a positive bias. This solution reduces the flow gain when the cross-sectional area approaches zero. It is, for instance, used for the flight control actuators of certain helicopters in order to improve the rejection of mechanical vibrations introduced by the linkage system transmitting the pilot's orders or by the anchorage of the actuator on the helicopter frame.

For a *dual slope* restriction, the geometric characteristic curve has two separate linear domains (bilinear): a gentle slope followed by a steeper slope. In actuators with a position servo-loop, this solution is used to reconcile two requirements: first, stability at low speeds (gentle slope leading to a low loop gain); second, dynamic accuracy at high speeds (steep slope to increase loop gain and minimize tracking error). For the latter effect, stability is ensured by the fact that the flow-pressure gain decreases as flow increases, as dictated by the root-curved characteristic line of equation [5.14], [MAR 02b]. This restriction type is, for example, implemented in rudder actuators on Airbus A380 [ATT 08].

While associating variable restrictions together to perform the metering, the geometric characteristic curve choice for each orifice makes it possible to act on the overall values of the pressure gain, the flow gain and the leakage flow of the metering function. For a valve intended to ensure servo-loops, the geometric characteristic curves of the four variable orifices (a–d on Figure 5.11) are often the same. It is, however, possible to pick different curves for them, such as in the two specific cases that follow:

– different gains have been adopted for the channels associated to control ports C_1 and C_2. This is done in order to compensate for the asymmetry of hydrostatic cross-sections of the hydraulic load (differential cylinder) or for forces it has to apply (non-zero average force to apply);

– a more significant underlap is implemented on reservoir return channels (b and d in Figure 5.11). The purpose of this is to lower the pressure of hydraulic load chambers when the valve is in a centered configuration. This has a positive impact on the leakage flow and the hydraulic load lifespan. Figure 5.18 shows an example of such a set-up for a rudder servovalve of a large passenger jet [ATT 08].

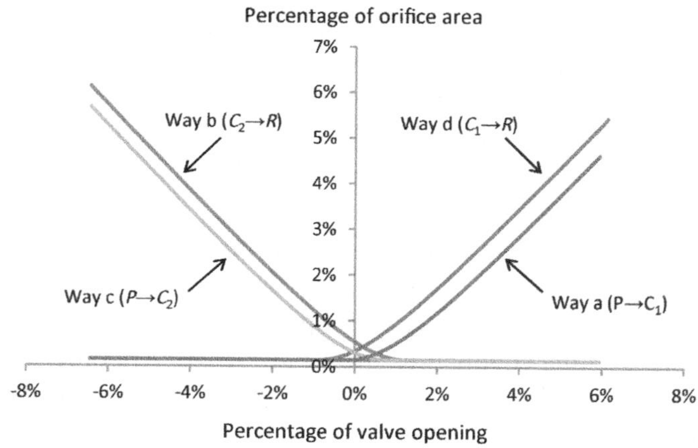

Figure 5.18. *Geometrical characteristic curves of the channels of a rudder actuator servovalve*

5.4.2.3. *Power capabilities*

This section is focused on the common case of the critical-closed-center, symmetrical, four-way valve with a linear geometric characteristic curve, supplied at constant pressure. Results are plotted in Figure 5.19. Other cases can be developed following the same pattern.

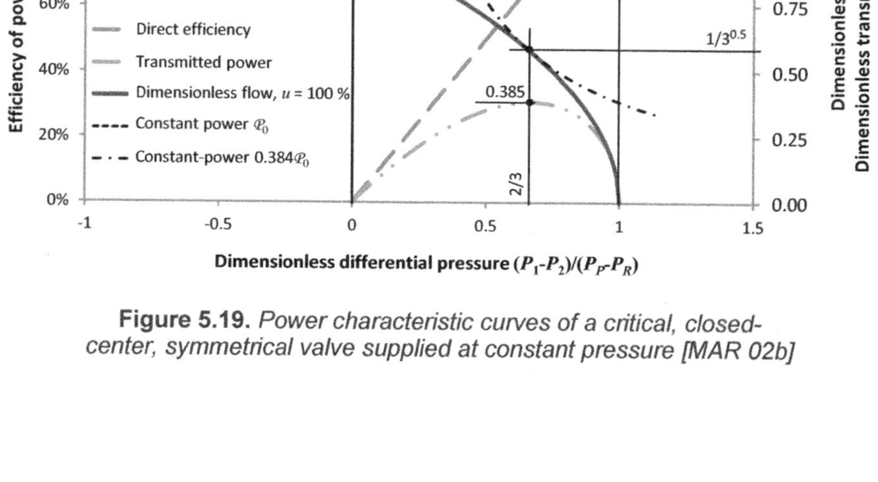

Figure 5.19. *Power characteristic curves of a critical, closed-center, symmetrical valve supplied at constant pressure [MAR 02b]*

The power transmission capacity of a metering hydraulic function is commonly characterized by the nominal flow Q_n. This flow is defined as the flow supplied by the valve, when the valve is subjected to a control signal, which causes it to fully open and when the flow generates, upon crossing this opening, a pressure drop equal to the nominal value[3] ΔP_n. The nominal pressure drop is usually set at $\Delta P_n = 70$ bars[4], in other words, 35 bars per orifice back and forth. For a critical, closed-center, symmetrical, four-way valve, the flow Q transmitted to the load depends on three quantities: the opening control signal u, the differential pressure applied to the load $\Delta P_{12} = P_1 - P_2$ and the working supply pressure $\Delta P_{PR} = P_P - P_R$. In order for the valve to supply flow to the load, hydraulic restrictions of active channels have to be subjected to a pressure difference. Indeed, this is equivalent to the behavior of an electrical resistor, which has to be subjected to a voltage difference in order to allow the current to pass through. The pressure difference ΔP_d at the valve is drawn from the working supply pressure ΔP_{PR}. Therefore, it is not available to the load. This is proven by the Wheatstone bridge in Figure 5.11:

$$\Delta P_{PR} = \Delta P_{12} + \Delta P_d \qquad [5.15]$$

Combining equations applied to each flow channel leads to an expression for the mean flow supplied to the load:

$$Q = u\frac{Q_n}{u_n\sqrt{\Delta P_n}}\sqrt{\Delta P_d} = u\frac{Q_n}{u_n\sqrt{\Delta P_n}}\sqrt{\Delta P_{PR} - \Delta P_{12}\,sign(u)} \qquad [5.16]$$

This characteristic is proportional to the opening control signal. It is also symmetrical in the (u, Q) plane. If the load is driving, then ΔP_{12} and u have opposite signs and the term under the square root is greater than unity. Therefore, the load forces flow when it is driving.

The no-load flow Q_0, which is supplied to the load at zero differential pressure ΔP_{12} for a wide open valve can be expressed as:

$$Q_0 = Q_n\sqrt{\Delta P_{PR} / \Delta P_n} \qquad [5.17]$$

3 In order to avoid any confusion with nominal user needs, the term "reference" should be used instead of "nominal".

4 This value was chosen because it corresponds to the pressure lost in the valve supplied at 207 bars when it transmits as much power as possible to the load.

In order to transmit power to the load, the latter has to be supplied in flow Q and in differential pressure ΔP_{12}. However, this is only possible if part of the working supply pressure is lost in the valve to the load. This is, once again, the exact principle of metering by a variable hydraulic resistance that was previously addressed. Here, apparent power $\mathcal{P}_a$ is introduced. It represents the power that the valve would transmit to the load if it could apply both the working supply pressure ΔP_{PR} and the no-load flow Q_0 to the load. Apparent power can be expressed as:

$$\mathcal{P}_a = Q_0 \ \Delta P_{PR} \tag{5.18}$$

For the previously stated reason, some supply pressure should be allocated to the valve to enable it to supply flow to the load. It can easily be proven from the previous equations that the valve transmits maximum power to the load when it spends a third of the working supply pressure. Hence, the maximum pressure actually transmitted is given by:

$$p_M = \frac{2}{2\sqrt{3}} P_a = 0.385\, p_a \text{ for } \Delta P_d = \Delta P_{PR} / 3 \tag{5.19}$$

Metering efficiency η_d represents the ratio between power supplied to the hydraulic load and power drawn from the hydraulic power source. Given that there is very low leakage at the valve (see Figure 5.20) metering efficiency is directly proportional to the ratio of load and supply pressures (except at a very small opening because of leakage):

$$\eta_d = \Delta P_{PR} / \Delta P_n \tag{5.20}$$

At the point of maximum power transfer, this efficiency equals 66%. It tends to be 100% as pressure drop at the valve tends to zero. However, in these conditions no power is transmitted.

The internal leakage flow Q_f of the metering function is equally important. Indeed, it has an impact on the sizing of distribution networks and hydraulic generation. It is generally measured by recording the return flow Q_R as a function of valve opening, at zero load flow ($Q_1 = Q_2 = 0$). Figure 5.20 gives an example of such a plot for a commercial airplane rudder actuator. Further study of the diagram on the right side of Figure 5.11 explains why the curve is bell-shaped and why it reaches its maximum when valve opening is zero. Indeed, cross-sectional areas being very small, leakage develops essentially in laminar flow. This explains its significant sensitivity to temperature: the maximal leakage flow is increased threefold between −40°C and 32°C (fluid viscosity typically gets divided by 6 between these temperatures). Nevertheless, maximum leakage flow remains very low since it only represents 0.3% of the nominal valve flow. Despite this low relative value and given that it is maximum at zero opening, the leakage flow penalizes positioning applications, which will make it work in the vicinity of this point of zero opening. Leakage flow, like flow and pressure gains (introduced in the next section), is very heavily affected by the clearance between moving parts of the valve and by radius of curvature of orifice edges. This radius depends not only on the quality of manufacture but also on the wear caused by erosion under the effect of high fluid velocities in variable sections. Therefore, radius of curvature increases with the duration of use. This is why the health status of a hydraulic system can be monitored via its level of internal leakage.

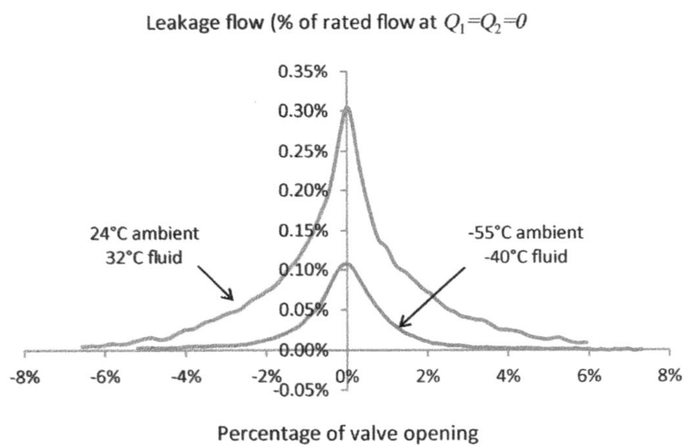

Figure 5.20. *Leakage flow of a rudder actuator valve [ATT 08]*

The power characteristic curve in the force-speed plane of an actuator involving this type of valve and a hydromechanical power transformer is obtained by combining equations [4.1] or [4.2] with [5.16]. In order to get realistic results, it is necessary to take into account internal leakage between lines 1 and 2 as well as pressure drops on lines connecting the valve to the hydraulic load. The aim is to minimize the diameter of these lines in order to minimize the weight and overall dimensions of the actuator. Therefore, these pressure drops are often far from negligible.

5.4.2.4. *Properties with respect to control*

Regarding controls, the dynamics, stability and accuracy of the valve opening in response to an opening control signal are paramount. Beyond these common quantities, the static hydraulic characteristic curve of the metering function is of particular relevance. This is because it heavily impacts actuation performances [ATT 08, MAR 02b]. For four-way valves (see Figure 5.11 left), three gains are commonly used to characterize the relationship linking the opening control signal u, the flow Q and the differential pressure ΔP_{12}. Figure 5.20 gives an example of flow and pressure gains for the same actuator servovalve as the one used in Figure 5.18 [ATT 08].

The flow gain represents the valve opening dependency of the flow Q supplied to the load, at constant differential pressure ΔP_{12}. Condition $\Delta P_{12} = 0$ is the most interesting case. From a functional point of view, this gain is used when the valve opening is considered to act on the flow while pressure is considered a disturbance. For actuators with a position servo-loop, this gain contributes directly to the loop gain. The latter is used for the study of dynamics and stability. In Figure 5.21, the functional dual-slope shape can clearly be identified. The slope changes for a relative valve opening of 25%. Moreover, it can be noted that the overlap at low temperatures is very low. On a side note, the offset between the two plots is due to the valve control function and the opening measurement.

The pressure gain represents the valve opening dependency of the differential pressure ΔP_{12}, at given control flow Q. Condition $Q = 0$ is the most interesting case. From a functional point of view, this gain is used when the valve opening is considered to act on the differential pressure while flow is considered a disturbance. For instance, this gain is used to assess the position error under load of hydraulic position servo-loops and in the absence of leakage. Figure 5.21 exhibits a slight bias (0.2–0.3%) at the hydraulic zero, which is defined by $P_1 = P_2$. It also shows that at hydraulic zero, pressures are lower than the net supply half-pressure. This reflects a favorable return underlap. Finally, it is interesting to see that the pressure gain in the vicinity of the hydraulic zero is greater than 50% of the net supply pressure for a 1% valve opening. This high gain significantly contributes to the accuracy of position servo-loops with closed-center valves, since it enables the development of significant hydrostatic force for a weak control signal. This makes it possible to set the load in motion or to maintain it at a standstill despite the presence of significant external force.

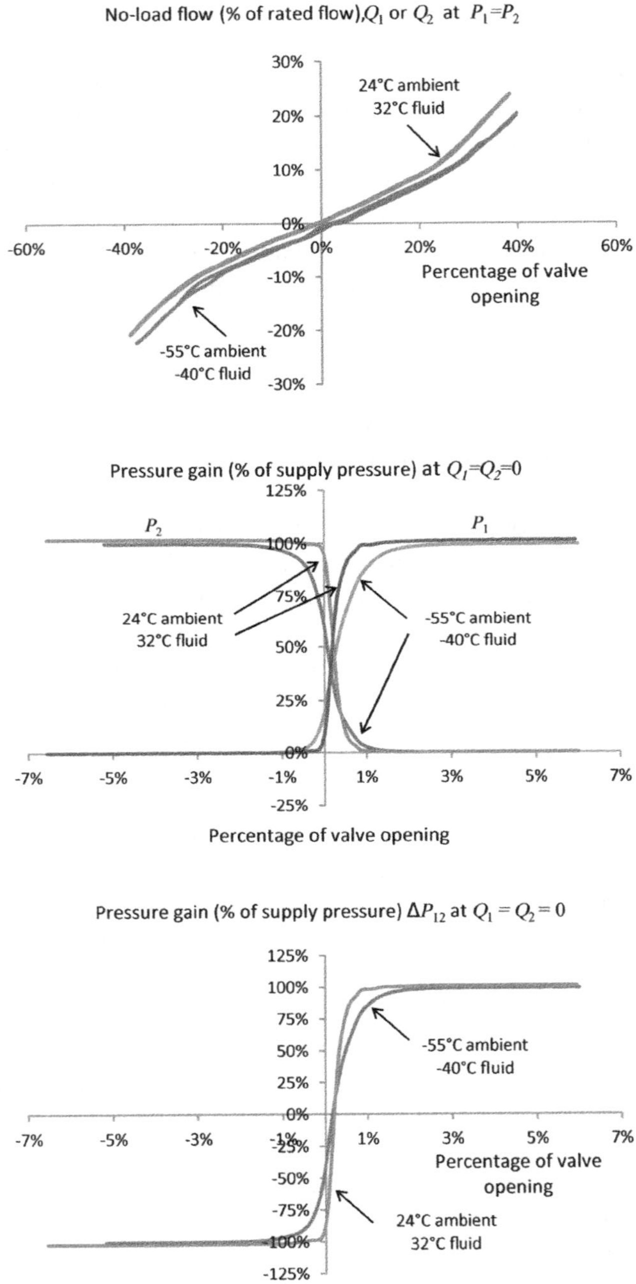

Figure 5.21. *Static hydraulic characteristic curve of a rudder actuator servovalve [ATT 08]*

The pressure-flow gain for a given valve opening gives the hydraulic resistance characteristic curve. This gain ties the differential pressure P_{12} together with the control flow Q for a given opening. It makes it possible to take into account the flow-pressure coupling introduced by the restriction. This coupling generally improves the stability at the expense of error under load [MAR 02b].

5.4.2.5. *Interfacing and valve opening controls*

Valve opening controls can be mechanical. For example, this is the case for hydromechanical actuators, which are found on most helicopters and which make up the majority of actuators on commercial airplanes pre-Airbus A320. The controls can be interfaced within the valve with electrical control signals. This is the case in servovalves, which are introduced in section 5.5.

In most applications, for instance flight controls or landing gear steering, power metering must enable the actuator to be operated in a position servo-loop. For this purpose, a "flow control" valve is used. Ideally, it should deliver flow according to its opening setpoint. However, in practice there is no easy way to construct such a valve that would require flow measurements. This is why so-called "flow control" valves instead modulate their opening in accordance with the control signal they receive. Load pressure ΔP_{12} then appears as a disturbance since it impacts the flow exchanged with the load, as can be seen in equation [5.16] and Figure 5.19.

In other applications, such as braking or controlling the displacement of a hydraulic generation pump, the aim is to control pressure. For this purpose, a "pressure control" valve is used. This valve is internally set up in such a way that its opening makes the pressure it applies to the load proportional to its control signal. Unlike the previous case, here, the flow supplied to the load is what appears as a disturbance for the metering function.

Figure 5.22 gives examples of symbol depictions for different types of controls. The first valve corresponds to a hydromechanical control, the second to an electro-hydraulic control with servovalve, the third to a pedal braking control and the fourth to an electrical braking control.

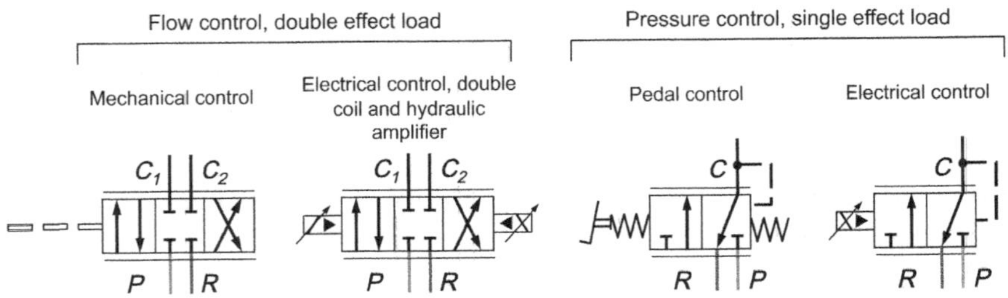

Figure 5.22. *Different types of controls for hydraulic valves*

In order to maintain or modify the opening of a variable restriction, actuation forces have to be developed between the moving part and the body of the restriction [GUI 72, FAI 81]. These forces have several different components:

– Fluid/solid *viscous friction forces* generated by the fluid flowing in gaps parallel to the direction of travel of moving parts. These forces are generally negligible, even at low temperatures. For example, 13N maximum for a cylindrical spool of a diameter 6.35 mm with 4 lands of 10 mm and a spool-sleeve radial clearance of 2 µm, when it has a sinusoidal movement of amplitude 0.25 mm at 50 Hz, for a fluid with specific density 850 kg/m^3 and viscosity 500 cSt.

– Solid/solid *stiction forces* originating from the rupture of the fluid film between the bearings of fixed and moving parts, under the influence of transverse forces or imperfect geometries. These forces can potentially turn out to be significant but this is not worrying because there are well-known solutions to greatly reduce them. For linear cylindrical valves, balance grooves are manufactured on spool bearings [SWE 51]. They are just about distinguishable in Figure 5.12. For rotary cylindrical valves, the spool is mechanically centered within its sleeve with the help of needles. For example, this is the case on actuators of the main rotor of helicopter NH90.

– Parasitic *hydrostatic forces* due to the summation of pressure forces in the direction of the valve opening. These forces cancel each other out by design in sliding and symmetrical rotary restrictions because establishing hydrostatic equilibrium is easy.

– *Hydrodynamic forces* originating from the change in the fluid as they flow through the valve [FAI 81]. When aiming to change directions (stationary hydrodynamic forces) or accelerate (transient hydrodynamic forces), the fluid "bears against" the restriction walls. Transient hydrodynamic forces are generally weak for common applications. Nevertheless, under a constant pressure drop at the valve, steady-state hydrodynamic forces resemble elastic return forces pulling toward zero opening. Similarly to hydrostatic forces, steady-state hydrodynamic forces, therefore, require the application of a permanent force in order to maintain a given valve opening. These forces tend to close the valve. They are often the most significant contributors to the overall force that has to be generated in order to actuate the moving parts of variable restrictions. Certain specific geometries that have been known for more than 50 years [CLA 57] make it possible to partially reduce steady-state hydrostatic forces in linear cylindrical valves. For the same spool as previously, calculations show that the overall stationary hydrodynamic force without compensation is equal to 55 N at full opening, for a no-load flow of 35 l/min, under a net supply pressure of 207 bars.

– *Inertial forces* that are associated with the acceleration of restriction moving parts. For the majority of applications, these forces are not very significant. This is because the mass of moving parts (from a few tens up to a few hundred grams) and their strokes (from 0.1 mm up to a few mm) are small. However, in the case of highly dynamic applications, inertial forces can become non negligible[5]. By way of example, the inertial force calculated equals to 1.68 N if the same spool as previously, which has a mass of 20 g, responds to a position change of 0.25 mm in 5 ms as a second order system with a dimensionless damping factor of 0.7.

– *Shear forces* which can appear if a solid particle accidentally hampers restriction closure. It can be calculated that 200 N are needed to cut thought a steel particle that is highly resistant to shear (500 MPa) of cross-section 0.4 mm², equivalent to a diameter of 0.71 mm. This "accidental" force often calls for much greater control force for the variable restriction. Fortunately, this force capacity is only required in transient and abnormal situations. Furthermore, the steady-state hydrodynamic flow force helps close the valve because it acts in the same direction as the force to shear does.

5 When the moving part of the valve is controlled mechanically, such as for example in a direct control valve, inertial effects reflected by the mechanical control device should not be neglected.

5.4.2.6. *Resistance or tolerance to jamming*

Jamming (or seizure) of a metering valve in the open configuration is a serious event, which can have catastrophic consequences. Indeed, in the event of a seizure, the valve connects the hydraulic load to the power source and the reservoir return through hydraulic resistances that have become fixed. The load then starts moving with uncontrollable speed until it finds mechanical equilibrium, for instance when in reaches the end-stop. The occurrence probability of this event can be greatly reduced by design: the metering function can be given the ability to resist jamming.

First of all, the resistance to jamming of a metering hydraulic valve is improved by adding a “last chance” or an “integral” filtering element within the valve. Sufficient shear force to cut through the particle responsible for the seizure should also be made available. Thanks to these arrangements, the failure rate by jamming of the metering function is estimated at 10^{-7}/flight hour for servovalves [MIL 93]. However, in-service experience feedback collected shows that this value may be ten times greater, depending on applications.

Next, redundancy is put to use by nesting together two metering valves, as illustrated in Figure 5.23. From a hydraulics standpoint, the two valves are connected in parallel but from the standpoint of mechanical opening controls they are connected in series. In the normal mode, the opening of variable orifices follows from the movements of the main spool with respect to the main sleeve. In the absence of jamming, the main sleeve is held in place with respect to the valve body by a springbox (with calibrated springs). In the event that the primary valve gets jammed, the main sleeve starts acting as a backup spool: it moves with respect to the valve body, which now acts as the backup sleeve. This displacement can be detected by a sensor, which then sends a warning, alerting failure by the jamming that has occurred. Therefore, the resistance to jamming of the metering function relies on the jamming tolerance of the main valve associated to the springbox. For nearly 50 years, this solution has exhibited a jamming failure rate lower than 6.5 10^{-9}/ flight hour [NGU 04] for the rotary valves on the right of Figure 5.23. However, as with all passive systems in normal mode, it is necessary to safeguard against a dormant fault. This is done by a procedure or a set-up allowing the health status of the secondary valve to be checked periodically.

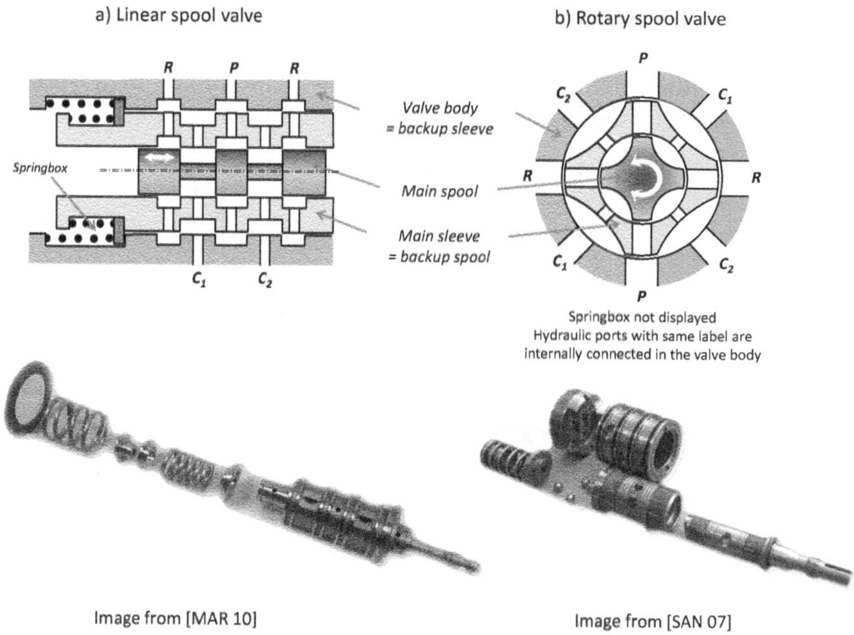

Figure 5.23. *Redundant power metering valves with linear cylindrical spool (left) and rotary spool (right)*

5.5. Servovalves

Over the past few decades, the transmission of information in the electrical form for hydraulic power actuators has become mainstream. The universalization of this technology was made possible thanks to the availability and maturity of proportional and dynamic electro-hydraulic power interfaces i.e. servovalves (Electro-Hydraulic ServoValve). Due to its specificity and its importance for aircraft actuation systems, the servovalve[6] is well worth addressing in a dedicated section.

5.5.1. *Architecture*

The servovalve assumes an electro-hydraulic interface function. This function consists of metering the power transmitted to the hydraulic load

6 The term "servovalve" can be interpreted as referring to either a valve used for servo controls or an internally servolooped valve; both definitions being acceptable.

from the hydraulic source, as a function of the electrical signal it receives. Figure 5.24 depicts the generic architectural design of a servovalve. Depending on the applications, either a “flow servovalve” or a “pressure servovalve” is used (see section 5.4.2.5).

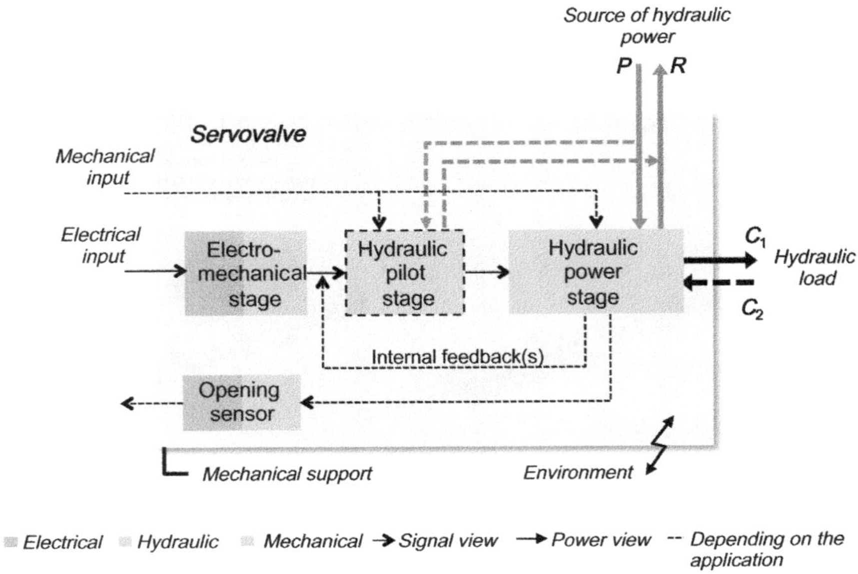

Figure 5.24. *Generic architecture of a servovalve*

Unfortunately, the technology to directly go from the electrical domain to the hydraulic domain does not exist. Therefore, an electromechanical stage (electric motor) has to be added: this allows it to first go to the mechanical domain from the electrical domain. As for hydraulic power metering, it is carried out by a hydraulic power stage. This stage performs parallel power distribution, with a permanent pressure source (P, R) and mechanically controlled variable hydraulic resistances (as depicted in Figure 5.11). Depending on the type of hydraulic load, the power stage supplies either one or two hydraulic ports (C_1, C_2).

Direct actuation by a Direct Drive Valve (DDV) of the hydraulic power stage by the electric motor calls for the ability to develop significant forces on the moving part of the power stage (see section 5.4.2.5). This need increases the motor weight and the power supplied to controls. This is why, in most applications, the servovalve includes a hydraulic pilot stage. This stage is powered by the hydraulic power source. It is intended to amplify forces developed by the motor in order to make them reach the required strength on the hydraulic power stage.

Figure 5.25 depicts, in the form of a network diagram, the "signal" view of the most common types of servovalves[7]. Curved arrows indicate interactions between elements that are considered disturbances in this view. They reflect the fact that elements not only exchange signals but also power and so the "power view" should not be neglected.

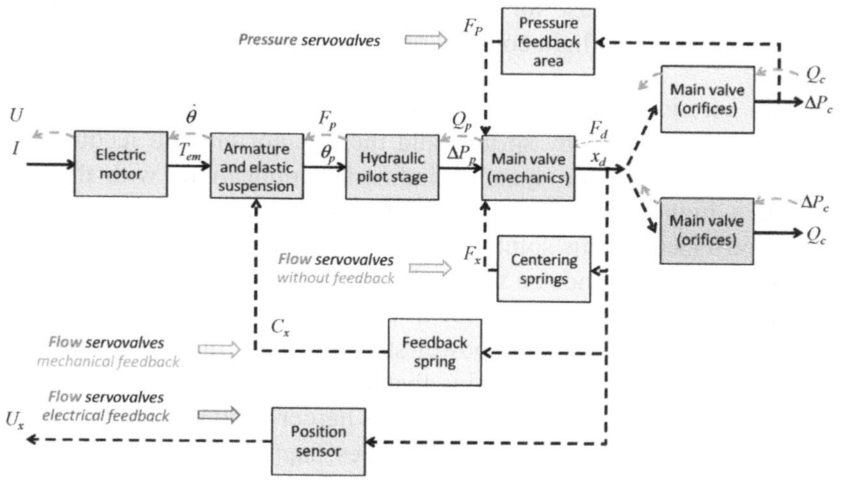

Figure 5.25. *Signal view of different types of servovalves*

From a functional point of view, the electromechanical stage generates electromagnetic torque T_{em} on the moving part of the pilot stage. In static mode, this torque is proportional to the control current I. The moving part (armature) is suspended elastically with respect to the servovalve body. It, therefore, adopts a position at an angle θ is defined by the overall torque that it is subjected to. The pilot stage creates a pilot differential pressure ΔP_p on the power stage, proportionally to its displacement. This differential pressure is applied to the piloting sections of this last stage in order to develop high driving forces (a few dozen daN) that are required for actuation. Hence, on the scale of the actuator, the electric motor control input is interpreted as a signal because it transmits very low power (a few dozen mW) compared to power delivered by the actuator (from a few kW up to a few dozen kW).

7 The meaning of variables used in the diagram are in accordance with the general nomenclature.

During the early days of servovalve use, the position loop of the actuator did not include any electrical elements yet, for reliability reasons. Back then, the loop was implemented by mechanical feedback. This was done by acting on the balance of the moving part of the servovalve electromechanical stage, as a function of the load position. The position servo-loop of the actuator could, therefore, only be corrected hydromechanically. It could, for example, be integrated in the servovalve to make correctors (Pressure-Flow, Dynamic Pressure Feedback and Static Load Error Washout) [TAY 65, GEY 72]. For instance, this was the case for thrust vector control of Saturn launchers and Space-Shuttle, and for flight controls of General Dynamics F111 bomber.

Within certain flow servovalves, the moving element of the hydraulic power stage is centered by springs and there is not internal feedback loop. This solution is used for dynamic industrial applications with the aim of obtaining a high bandwidth with no instability risk. However, the static accuracy provided by this solution is mediocre. Furthermore, the servovalve static gain heavily depends on the supply pressure, which influences the pilot stage gain. Flow servovalves selected in the aerospace are equipped with a feedback loop on the hydraulic power stage opening. Indeed, the balance of the moving part acts on this opening thanks to an elastic needle that converts position into a feedback torque. Stability is harder to adjust, but the feedback loop improves static accuracy and reduces supply pressure sensitivity: a change in the pilot stage gain mainly modifies the internal loop gain and hardly affects the servovalve static gain.

For pressure servovalves, the feedback loop is carried out by applying a force on the moving element of the power stage: pressure is applied on a secondary piloting section of this element. When this is the only loop present, the pilot stage operates in open-loop. This makes the servovalve static gain sensitive to supply pressure, as with flow servovalves with no internal feedback loop.

The power stages of certain servovalves are also equipped with a mechanical drive input. For example, this design allows electrical demands originating from the autopilot to be superimposed on the pilot's mechanical setpoint. This solution also makes it possible to manually takeover opening controls of a valve, in the event of a fault on the electrical chain. It is for instance implemented on Airbus H225M helicopters in the autopilot hydraulic block.

Finally, servovalves include more and more often a power valve opening sensor, which is essentially used to perform monitoring tasks[8].

5.5.2. *Incremental improvements of servovalve performances*

Table 5.3 provides the timeline of the history of servovalves, which was compiled based on data provided by [MAS 78]. Key innovations are highlighted in bold.

Date	Electromechanical stage	Hydraulic stages	Closed-loop operation	Remarks
1922	N/A	**Flapper-nozzle amplifier concept (pneumatic)**	N/A	Application to chemical processes with pneumatic control [BEN 93]
1938	N/A	**Jet-pipe amplifier concept (pneumatic)**	N/A	Used for computers and fluidic controls [WUN 38]
1946	Proportional linear electromagnet	**Spool pilot stage + power stage**	No, spring centering	Power stage driven by large forces. Reduced hysteresis and sensitivity to surroundings [TIN 49]
1950	Electromagnetic torque motor	**Single flapper-nozzle pilot stage** Spool power stage	No, spring centering	High bandwidth and reduced threshold [WIL 50]
1951	**Symmetrical electromagnetic torque motor**	Single flapper-nozzle pilot stage Spool power stage	No, spring centering	Reduced temperature drift [WIL 53]

8 Certain industrial servovalves replace the needle mechanical feedback loop by an electrical servo-loop that relies on the opening information provided by this sensor. In this case, the electronic card that performs this servo-loop is most often integrated in the servovalve (sensor conditioning, control signal construction, electrical amplification).

1953	Symmetrical electromagnetic torque motor	Single flapper-nozzle pilot stage Spool power stage	**Power spool with mechanical position servo-loop**	Improved static accuracy by mechanical feedback of the power spool on the torque motor [CAR 53]
1955	**Dry electromagnetic torque motor**	**Double flapper-nozzle pilot stage** Spool power stage	No	Reduced sensitivity to solid pollution of the fluid [BOY 55]
1957	Symmetrical electromagnetic torque motor	**Jet-pipe pilot stage** Spool power stage	Power spool with mechanical position servo-loop	First stage solid pollution leads to passive failure instead of active failure [ATC 57]
1967	Symmetrical electromagnetic torque motor	**Jet deflector pilot stage** Spool power stage	Power spool with mechanical position servo-loop	Alternate solution to jet-pipe with same benefits for the first stage solid pollution [MCF 67]

Table 5.3. *Timeline of the historic development of servovalves*

Nowadays, the basic principles of servovalves have not evolved much since the late 1960s. Servovalves now offer great performances and a high level of maturity, which stems directly from past developments:

– temperature drift is strictly limited by the design of the different stages, which are all made geometrically symmetrical;

– static accuracy (threshold, hysteresis, sensitivity to hydrodynamic forces and parasitic accelerations) is improved by the implementation of an internal position or pressure feedback loop and by the use of contactless pilot stages. Indeed, the latter make it possible to develop significant driving forces without introducing static nonlinearities;

– resistance to solid pollution is improved by four design specifications: isolating the electric motor from the hydraulic fluid (dry motor), developing large driving forces allowing us to cut through a solid particle in the orifices of the power stage, using pilot stages of the jet-pipe or jet deflector types whose pollution can only lead to passive failure and finally by adding an integral filter.

5.5.3. *Power supply of the electromagnetic motor*

Torque motors have at least two (or four) distinct coils. These windings can be powered by two (four) distinct control channels or they can be connected in series, in parallel or in push-pull. The parallel configuration enables static redundancy to be implemented with regard to the rupture of a winding or of one of its inherent connections.

As mentioned previously, the opening of the hydraulic power stage depends on the current supplied to the electric motor. From this point of view, it is, therefore, interesting to control the servovalve in terms of current. The low pass filter effect that would be introduced by voltage control is thereby avoided. Indeed, if controlled in terms of voltage, the series association of the resistance R with the inductance L of the coils would establish a first order dynamic of electromagnetic time constant $\tau_{em} = L/R$ between the voltage and current.

5.5.4. *Concepts of pilot stages*

For the purpose of constructing a hydraulic pilot stage, a cylindrical spool four-way valve, such as the one shown in Figure 5.11, can be used. For example, this solution is implemented by SABCA in thrust vector control actuators of solid boosters of European launcher Ariane 5 (see Figure 5.26 right). Each actuator is equipped with two pilot valves operating in active/passive mode in order to drive a high power hydraulic valve (at a supply pressure of 380 bars, each actuator develops a force of 347 kN and a power of 132 kW with a bandwidth of 7.9 Hz). In this design, the electric motor and the spool pilot stage are bundled together as a unique component. The cross-section of this component is shown on the left of Figure 5.26. The electric motor axis is perpendicular to the hydraulic spool axis. Additionally, the conversion of translational motion into rotational motion is carried out by an eccentric effect. The spool is spring-centered. The power valve spool is position-servolooped by the opening control signal. Such a servoloop is carried out by acting on the motor power supply, according to a state feedback control structure.

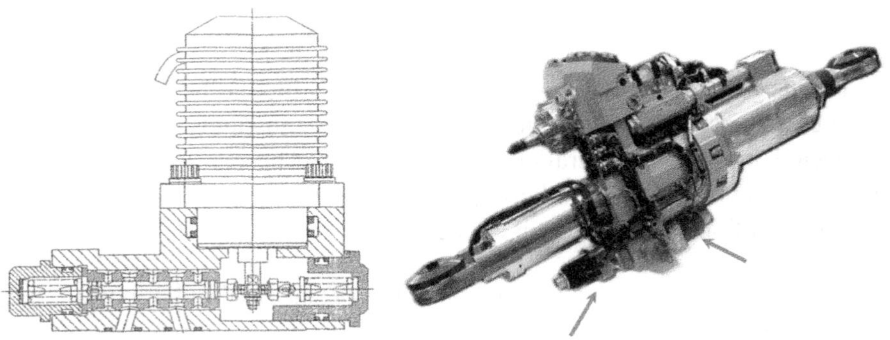

Figure 5.26. *Cross-section of the proportional SABCA pilot valve and its implantation on the thrust vector control actuator of launcher Ariane 5*

However, the existence of parasitic stiction forces and the elevated pressure gain of spool stages are the reasons why this solution is not recommended to construct direct drive valves (DDV) that are intended to supply "only" a few dozen liters per minute. The so-called "pressure" or "flow dividing" stages, like the ones depicted in Figure 5.27, are preferred. They have two main benefits for a piloting function. First is the fact that they offer the possibility of elastically suspending the moving parts. This removes dry friction and play issues. Second is the progressiveness of the differential pressure created at control orifices as a function of moving part displacement.

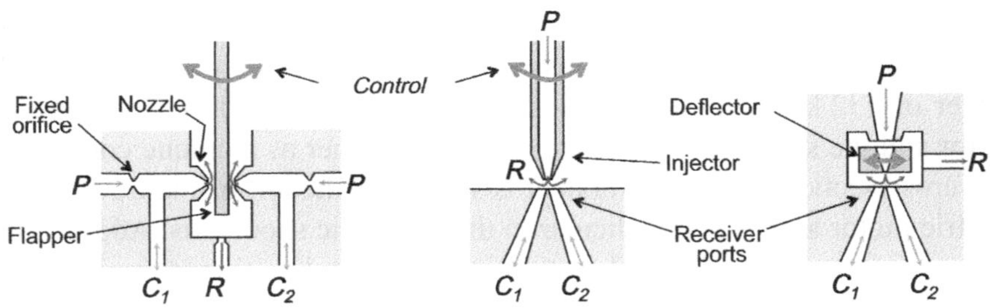

a) Double flapper-nozzle b) Jet-pipe c) Jet deflector

Figure 5.27. *Concepts of hydraulic pilot stages. For a color version of the figure, see www.iste.co.uk/mare/aerospace1.zip*

The *double flapper-nozzle* type solution, which can be found on the left of Figure 5.29, forms a complete (four-way) Wheatstone bridge. Two hydraulic resistances are fixed (fixed orifices) and two others vary in opposite ways at the nozzles. An additional hydraulic resistance is generally inserted in series on the return line in the form of a fixed orifice. Its purpose is to reduce the return pressure sensitivity as well as the risks of desorption and cavitation. The flapper is rigidly connected to the electric motor armature and is held by an elastic pivot with respect to the servovalve body. The double flapper-nozzle design is relatively well mastered because it is easy to model. On the other hand, this structure is sensitive to pollution in two ways. At neutral, the nozzle-flapper distance is typically of 30 to 40 μm, the diameter of fixed orifices is typically of 125 to 250 μm and nozzle diameter ranges between 250 and 500 μm. Hence, a solid particle of a few dozen micrometers can give rise to a pilot stage fault. Furthermore, the partial blockage of a fixed or variable orifice of the bridge by a solid particle results in active failure. Indeed, in these conditions, the pilot differential pressure at control orifices takes a non-zero value in the absence of a control signal. This causes the power stage to open and results in high speed movement of the hydraulic load it supplies. As a consequence, in aerospace, this solution has been nearly entirely abandoned.

The jet-pipe type solution, which is depicted in the center of Figure 5.29, is based on the conversion of kinetic energy at the nozzle jet into pressure energy at the receiver ports. The injector is rigidly connected to the electric motor armature and is held by an elastic pivot with respect to the servovalve body. The conversion of kinetic energy into hydrostatic energy is difficult to model accurately and implementing a jet-pipe stage is tricky. One of the two major benefits of this solution is its resistance to fluid pollution. Indeed, first, minimum fluid path cross-sectional areas are larger than in the flapper-nozzle solution: the inner diameter of the injector typically ranges between 200 and 500 μm, the diameter of receiver ports is twice that and clearance between the injector and the receivers equals several times this diameter. Second, solid pollution that could come clog the injector causes a passive failure: following clogging, fluid velocity decreases and the geometry of the jet changes. This reduces the gain and introduces a slight bias on the force amplification function. The second key benefit of this solution regards the permanent leakage flow. Indeed, for identical drive characteristics, leakage flow of the jet-pipe is significantly lower than that of the flapper-nozzle. These advantages explain the universalization of jet-pipe type pilot stages in aerospace.

The jet deflector is an alternative to the jet-pipe solution, which was introduced by a competing manufacturer. Both solutions offer the same benefits with regard to fluid solid pollution. Here, the injector is fixed. Therefore, it is the displacement of a deflector connected to the motor moving part, which steers the jet in order to favor a receiver port.

The static hydraulic characteristic curves of Figure 5.28 show the good progressiveness and symmetry of the pressure as a function of deflection, for the flapper-nozzle and the jet-pipe solutions. In order to make it dimensionless, the pilot differential pressure $\Delta P_p = P_1 - P_2$ is expressed as the ratio to the supply pressure. Similarly, the mechanical input deflection (flapper or injector) is expressed as the ratio to the reference deflection, which corresponds to the flapper gap at neutral or to the injector orifice diameter. The decreased flow Q^* supplied to the power stage is expressed relative to flow Q_0 when the nozzle or the injector are centered.

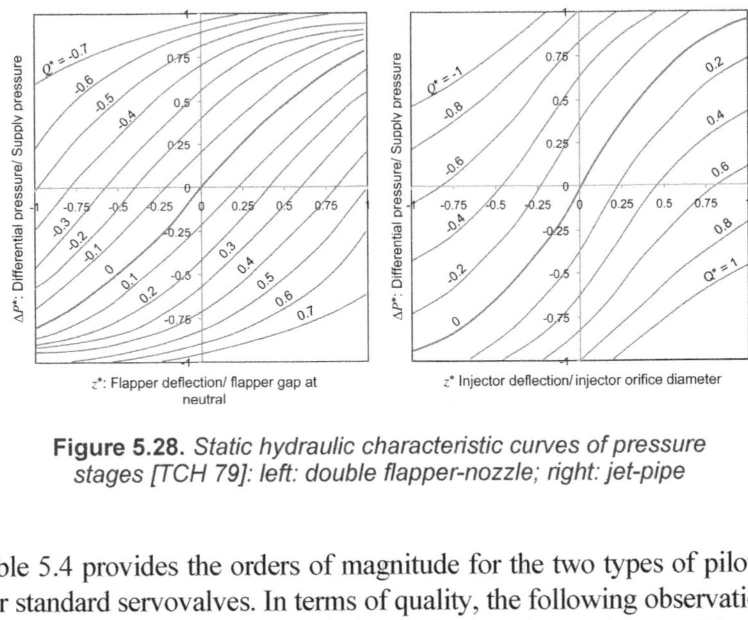

Figure 5.28. *Static hydraulic characteristic curves of pressure stages [TCH 79]: left: double flapper-nozzle; right: jet-pipe*

Table 5.4 provides the orders of magnitude for the two types of pilot stages and for standard servovalves. In terms of quality, the following observations can be made based on this table: control electrical power is very low, which makes it possible to reach a very high power amplification gain. The weight is reduced, which leads to an excellent power metering density, driving forces applied to the power stage are quite large and the frequency response is very satisfying. The vast majority of the leakage at neutral originates from the power stage and it decreases quickly with its opening. However, pilot stage leakage is permanent and typically represents from 0.6 up to 0.76% of maximum flow supplied by the servovalve. The size of the pilot stage internal protection filter confirms the low sensitivity of the jet-pipe solution to pollution. Finally, the fact that the spool stroke is very small illustrates the level of accuracy required for manufacturing. Indeed, in order to have control over the overlap, spool and sleeve active edges have to be relatively positioned with an error of less than 1% of the nominal stroke, i.e. just a few micrometers.

	Moog Type 30 (31–35)	Abex/Parker (415–450)
According to commercial documentation		
Pilot stage concept	Double flapper-nozzle	Jet-pipe
Flow rate at 35 bars per orifice (l/min)	26–379	18–265
Control electrical power (mW)	50	32–76
Mass (kg)	0.19–0.97	0.35–8.61
Driving force applied to the power stage (N)	245–712	140–2,200
Power stage spool stroke (mm)	0.250–0.500	0.304–1.447
Frequency for a −90° phase at a net pressure supply of 207 bars	150–47 (input ± 25%)	130–40 (input ± 50%)
Permanent leakage at neutral (l/min)	2.15–20.41	0.68–9.46
First stage supply filter (μm absolute)	30	90–250
Values calculated from manufacturers data at a net pressure supply of 207 bars		
Maximum hydraulic power (kW)	6–87	4.35–61
Power metering density (kW/kg)	31–89	12.3–7.1
Maximum power control gain (kW/mW)	0.12–1.74	0.13–0.80
Leakage expressed as % of maximum flow supplied	4.76–3.11	2.08–2.06
portion due to pilot stage (%)	0.1–0.76	?
Maximum estimated hydrodynamic force with no compensation (N)	72–1,047	52–733
Maximum ratio between estimated hydrodynamic force and driving force	3.4–1.1	2.7–3

Table 5.4. *Orders of magnitude for standard servovalves*

Figure 5.29 schematically illustrates the design of servovalves with flapper-nozzle or jet-pipe type pilot stages. It is worth noting that the electromagnetic torque motor is “dry”, meaning it is not bathed by the hydraulic fluid thanks to the pilot stage design (flexure sleeve for instance).

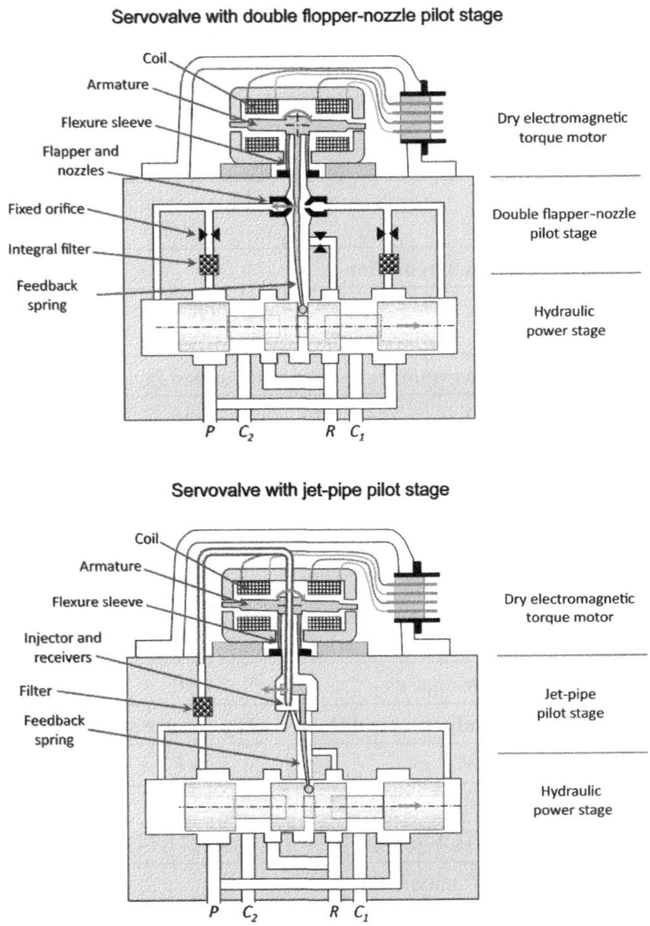

Figure 5.29. *Schematic of a servovalve with flapper-nozzle or jet-pipe pilot stage*[9]. *For a color version of the figure, see www.iste.co.uk/mare/aerospace1.zip*

Despite their many benefits, hydraulic pilot stages also have disadvantages that are less and less accepted for a servovalve:

– the pilot stage gain is defined as the slope of the static characteristic curve linking pilot pressure to deflection. As can be seen in Figure 5.28, it directly depends on supply pressure (because of its effect on the nozzle flow or on the injector flow at neutral) and it decreases with the flow supplied to

9 The structure of flexure springs has been modified here for the sake of the planar representation. In practice, they are constructed differently.

the power stage. Consequently, the bandwidth and the accuracy of flow servovalves are reduced for high magnitudes of the control signal (see Figure 5.30) and if the supply pressure decreases. Generally speaking, this effect is not very restricting for commercial aerospace because the servovalve is not a limiting factor of actuator performances;

– the failure rate of a servovalve (depending on manufacturers λ ranges from 0.25 to 1.4 10^{-6}/FH) means it cannot be used in simplex configuration for critical functions. This failure rate is incompatible with this configuration partly because of pilot stages. In order to bring the failure rates down to values tolerated, multiple and complex redundancies have to be implemented [TAY 76]. However, such redundancies have a very negative impact on the weight breakdown of the actuation function;

– the operating principle of hydraulic pilot stages is based on the existence of a permanent flow between the pressure source and the return. However, this entails permanent energy loss. For example, for an aircraft equipped with 20 active flight control actuators, a consumption of 0.2 l/min at 350 bars at servovalve pilot stages leads to a permanent power loss of 2.33 kW. Furthermore, within its working range, the overall efficiency of a flow pilot stage is often lower than 10%;

– manufacturing servovalves is costly and it involves processes that have to be performed by hand and require a high level of skill;

– servovalves remain sensitive to pollution, which can accumulate in smaller orifices and in areas where flow never changes direction (injector and nozzle).

5.5.5. *Direct drive valve*

The removal of the pilot stage in favor of a direct drive is beneficial for several reasons. First, the absence of pilot stage leakage reduces, by as much as 60%, the overall hydraulic consumption of the servovalve. Next, the smooth operation of the DDV can be tested pre-flight with no hydraulic power. Moreover, it is possible to build a servovalve in the DDV version for high flows, which would have required a third hydraulic stage in the EHSV version. Finally, the DDV dynamic response no longer depends on the supply pressure or the control signal magnitude (except through motor sizing). However, concerning embedded applications, when it comes to ensuring good resistance to jamming for the power stage or large bandwidth for high flow rates, DDVs are far from the best solution because of the weight of the electric motor.

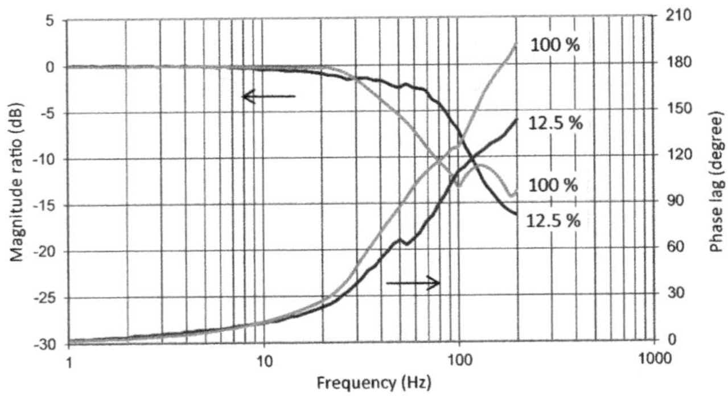

Figure 5.30. *Influence of the control current magnitude on the frequency response of a rudder actuator servovalve [ATT 08]*

5.5.5.1. *Applications*

DDVs have recently started being introduced on certain commercial airplanes on the braking backup channel. However, commercial aerospace still overwhelmingly relies on conventional servovalves with pilot stage for several key reasons:

– pilot stage leakage forces a permanent flow from the hydraulic power source to the actuator, and this is beneficial in certain conditions. Indeed, when flying at very low temperatures, actuators furthest away from the source (for instance ailerons and rudder), are naturally heated by the permanent flow of the first stage. This not only limits pressure drops in supply lines but it also restricts reduction of the servovalve power stages pressure gain;

– controlling servovalves in terms of current (a few mA) and at low impedance (a few hundred Ohms) naturally provides good immunity against electromagnetic disturbances. Control signals of a few mW can be directly emitted by computers without the need for power amplifiers. This is true despite the fact that servovalves are located several meters away from control electronics;

– the amount of electric power that has to be supplied to the electric motor of a DDV is typically a thousand times greater than the amount of power required by the electric motor of a two-stage servovalve (50 W compared to 50 mW). This has a significant impact on the architecture and the topology of the opening control signal transmission, which can no longer be viewed as simple information transmission.

On the other hand, DDVs have been introduced on several military aircrafts (Northrop B-2, Saab Jas39 Grippen, Dassault Rafale, Aermacchi M-346, AIDC-IDF) and helicopters (Eurocopter NH90) over the past few decades [HSU 91, MIL 93, SCH 93, NGU 04, SAN 07]. Using DDVs is most interesting for critical applications. In these situations, there is often only one distinct actuator per function, whose failure, therefore, has catastrophic consequences. When targeting identical reliability, response to failure and vulnerability properties, the overall weight breakdown of the actuator argues in favor of the DDV over the highly redundant EHSV, even though the DDV is heavier than the EHSV.

Main rotor actuator
NH industries NH90 [SAN 04]

Nose landing gear steering
Dassault Rafale [SAN 04]

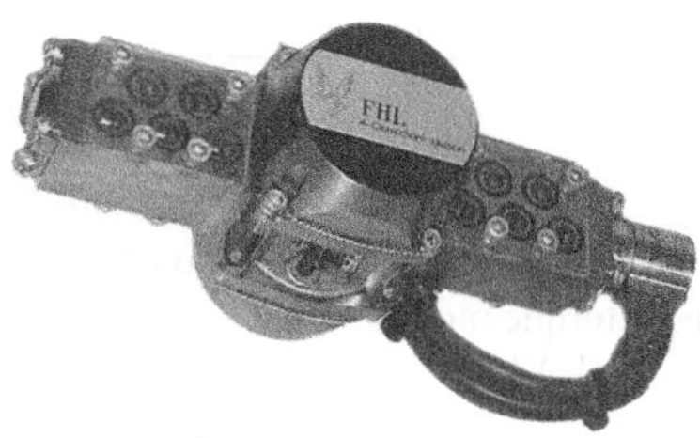

Leading edge actuation system
Eurofighter Typhoon

Wheel brake servovalve
Airbus A320

Figure 5.31. *Direct drive valves in service*

5.5.5.2. *Architecture of direct drive valves*

The amount of force needed to direct drive the hydraulic power stage is the key parameter that determines how the DDV electric motor is sized. Furthermore, the electromagnetic forces are generated by a motor result from the combination of magnetic field in the air gap and current in the windings. Both require heavy materials: magnets and iron for the magnetic field and electrical conductors for windings (copper wires). In order to use as little of these materials as possible (and consequently minimize motor weight), the amount of force it has to develop must be reduced. This can be achieved by:

– decreasing the force required to ensure DDV resistance to hydraulic power stage jamming;

– implementing the steady-state hydrodynamic forces reduction solutions;

– integrating a motion conversion function between the motor and the hydraulic power stage to reduce the amount of force at the motor (which will consequently have to develop a wider range of motion);

– avoiding the generation of transverse forces (especially by the motor or by the motion conversion element) because they favor stiction of the power stage.

These solutions can be found in the DDVs introduced in Table 5.5 and in Figures 5.31 and 5.32.

Manufacturer	Moog	Claverham	Goodrich
Aircraft	Fighter jet SAAB JAS39 Grippen	Fighter jet Eurofighter Typhoon	Helicopter NH industries NH90
Application	Primary flight controls	Secondary flight controls	Main rotor actuator
Failure rate		< 5.0 10^{-9}/flight hour	< 6.5 10^{-9} / flight hour
Consumption	Normal 0.45 A Shear 40 W / 1.25 A		
Electric motor	1 linear hybrid motor with 4 spring-centered coils	2 connected dry limited-angle torque motors (LAT)	4 limited-angle torque motors (LAT), 4×1.05 Nm 4 spring-loaded torque limiters
Motion transmission	Direct	Rotation/translation by an eccentric	Direct
Shear force	500 N	1,000 N	680 N
Valve	Simple linear	Linear tandem (± 1 mm)	Redundant rotary tandem
Position sensor	1 LVDT quadruplex	1 LVDT quadruplex	2 LVDT duplex

Table 5.5. *Architecture of three aeronautical DDVs in service*

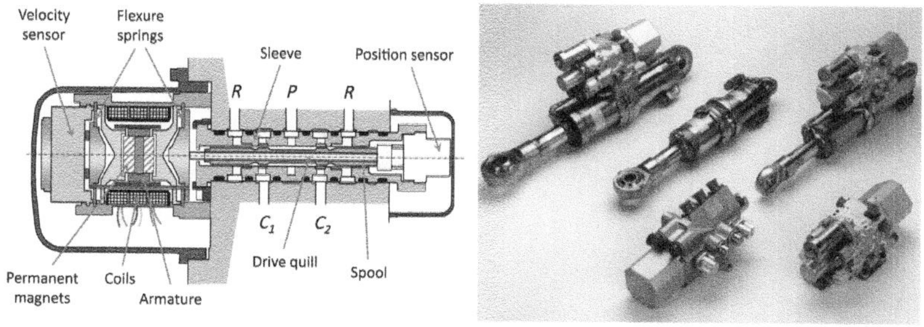

Figure 5.32. *Moog linear-motor direct drive valve: left: operating principle and components [SCH 93]; right: examples of DDV-equipped actuators (image © Moog). For a color version of the figure, see www.iste.co.uk/mare/aerospace1.zip*

6

Power Management in Hydraulics

6.1. Power distribution

The power distribution function consists of safely providing users with hydraulic power in whatever form they request, as a function of available sources and priorities. This function is implemented by associating, often dedicated, components and arranging them in generic configurations. The purpose of this chapter is to introduce and detail these arrangements and the components they rely on. The function/component matrix of Table 6.1 illustrates the relationship that exists between functions required for distribution and used components. The crucial function of hydraulic fluid conditioning was addressed in its specifically dedicated Chapter 3. In other respects, the distribution function is divided into several lesser functions (for instance, providing power, protecting and load managing) which themselves are divided into sub-functions. Seeing as this matrix is almost entirely diagonal, it can be concluded that most existing components are dedicated to a specific function.

6.2. Providing power

6.2.1. *Transporting fluid*

The fluid is transported between sources and hydraulic power users by lines that are interconnected to the equipment by fittings or connectors.

Component / Function and sub-function	Hydraulic lines, fittings and coupling	Shut-off valve and bypass	Shuttle valve	Flow divider	Priority valve	Accumulator	Pressure reducing valve/pressure booster	Bypass valve	Check valve	End-stop snubber	Pressure relief valve	Hydraulic fuse	Flow control valve
Providing power													
Transport fluid	●												
Connect	●												
Isolate		●											
Merge sources			●										
Share sources				●									
Prioritizesupply					●								
Store/restore energy						●							
Damp						●							
Adjust pressure level							●						
Load managing													
Lock in position		●											
Release (or de-clutch)		●											
Enable back drivability									●				
Damp										●			
Protecting													
Against under pressure									●				
Against over pressure/overload											●		
Against fluid over-consumption												●	
Against over speed													●

Table 6.1. *Function/component matrix*

6.2.1.1. *Hydraulic connectors*

Depending on the circumstances, hydraulic connectors can either be:

– permanent and non-disconnectable: in this case, they can be placed in areas inaccessible for inspection and maintenance;

– permanent and disconnectable: they enable the replacement of tubes and equipment during maintenance operations;

– temporary: they are then called couplers. Temporary connections are usually used in maintenance or ground operations (for example, power supplied by a ground power unit).

6.2.1.2. *Hydraulic lines*

Hydraulic line topology and routing are especially important in terms of eligibility, weight and maintenance. They will be addressed in detail in Chapter 7. Here, the remainder of this section focuses on selection criteria.

6.2.1.2.1. Nature of hydraulic lines

The selection of the type of line used is based on:

– The significance of relative motion between the ends of the line. If it is not negligible, such as in the case of supply lines for wheel braking or flight control actuators, flexible lines (also called hoses) are used. Conversely, if relative motion remains very small regardless of conditions, rigid lines (also called pipes) can be used. The reason for this is that only negligible mechanical constraints emerge from displacements in this situation and that pipes are more reliable than hoses.

– The area and the working pressure. For pipes operating at high pressures, titanium alloy is preferred. This is because its specific resistance (ratio between mechanical resistance and specific density) is typically 50% greater than that of stainless steel pipes. However, its low resistance to fire means it must be replaced by stainless steel for lines in concerned areas (engines or brakes for instance). Aluminum alloys, less resistant but lighter, are well suited for low-pressure areas. However, they too have to be replaced by stainless steel in areas prone to fire and projections (for instance, landing gear and landing gear bays).

6.2.1.2.2. Hydraulic line size

Pipe diameter selection is based on a trade-off that has to be considered over the entire working range, in particular with regard to the temperature range. Indeed, on the one hand, the diameter should be increased in order to reduce the average flow velocity v and, thus, minimize hydraulic resistance of the line to flow. On the other hand, the diameter should be decreased in order to minimize the weight of the fluid-filled line. In order to solve this dilemma in the best way possible, the additional constraints that define permissible fluid velocities are taken into account.

The first criterion concerns overpressure ΔP (or water hammer) in high-pressure lines due to the sudden closure of a valve. For example, this occurs when an active actuator switches to passive mode, following a fault detection. The overpressure originates from the conversion of the kinetic energy E_c of the fluid in the line into elastic energy E_e associated with fluid compressibility. The fluid kinetic energy is tied to the mass of fluid M and to its velocity v in the line. The elastic deformation energy is tied to the apparent Bulk modulus B_a of the fluid in the line, to the volume of fluid V and to the pressure change ΔP. Assuming that all kinetic energy before valve closure is converted into elastic energy after valve closure, the following expressions are obtained:

$$E_c = \frac{1}{2} M v^2 \qquad [6.1]$$

$$E_e = \frac{1}{2} \frac{V}{B_a} \Delta P^2 \qquad [6.2]$$

which can be rewritten as:

$$\Delta P = \sqrt{B_a \rho}\, v \qquad [6.3]$$

For example, for a commercial airplane fluid, the specific density ρ is close to 1,000 kg/m^3 and the apparent Bulk modulus B_a is of the order of 8,000 bars in a rigid line (pipe) at 207 bars. Considering a relative increase in pressure of 35% in the pressure line is tolerated, then the maximum permissible fluid velocity in the line before closure is $v_M = 8$ m/s.

The second criterion concerns the optimization of the hydraulic system overall weight. The weight of supply and return lines can only be reduced if the following two consequences are accepted: pressure drops caused by lines and weight increase for actuators. The net supply pressure of actuators also becomes decreased by this pressure drop as a consequence. For example, 20% of the net supply pressure can be allocated to pressure drop for power distribution. This pressure drop, first, results from resistance to flow which is evaluated by common distributed pressure loss models. Second, it follows from the fact that pressure is being spent to accelerate or slow down the flow in transient phases because of hydraulic inertia (see section 3.4.3).

Finally, regarding suction lines, the aim is to safeguard against the risks of fluid desorption or cavitation. Otherwise these phenomena would result in excessive pressure drop on the path between the reservoir and the pump.

For all these reasons, flow velocities ranging from 7 up to 10 m/s are adopted in high-pressure lines (depending on the working pressure), from 2 up to 4 m/s in return lines and from 1 to 2 m/s in suction lines. Recent studies [CAL 04] have shown that increasing fluid velocity above 10 m/s in high-pressure lines can pose potential risks regarding pressure surges and pressure pulses generated by the pumps.

Table 6.2 compares the performances of three standard diameter tubes of different nature for common flow velocities and for a phosphate-ester type IV fluid. Under an identical working pressure equal to 207 bars, compared to stainless steel, the titanium alloy decreases the mass per unit length of the fluid-filled tube by 35% per hydrostatic kilowatt transmitted at a full speed of 7 m/s. The mass ratio between the fluid and the dry tube typically drops from 1 to 0.5 between titanium alloy and stainless steel. Concerning return lines, tubes operate in steady state at low pressure, but they are selected to tolerate overpressures due to strong transient flows. This artificially penalizes the given values, as the hydrostatic pressure they convey is very low. It should be noted that filled tube weight only depends on the ability to transmit power via installed tubes. It is independent from the effectively transmitted power which is much lower on average, especially in cruising steady state. Pressure losses are calculated for a phosphate-ester type hydraulic fluid of grade 1. Compared to when temperature is average, it is clear that viscosity increase has a strong impact at very low temperatures and that it hardly has any impact at high temperatures. Overall, in order to transmit 60 kW of hydraulic power with a fluid at 38°C in a titanium pressure line and an aluminum return line, 1.5 kg of fluid and lines have to be setup on board per meter separating the source from the user. At full power capacity, resistance to flow in lines generates a loss of more than 100 W/m and an overall pressure drop (back and forth) of 0.36 bar/m. These values are almost doubled for a fluid at 15°C. Therefore, this example is a good illustration of the weight penalty caused by centralized hydraulic generation in conventional power architectures.

Nature of the tube	Operating conditions		Mass per unit length per kW (g/m/kW)	Power loss per unit length (W/m)		
	Pressure (bar)	Velocity (m/s)		–54°C	38°C	99°C
Titanium alloy 207 bars	207	7	14–16 depending on diameter	1,800	30–87 depending on diameter	22–64 depending on diameter
Stainless steel 207 bars	207	7	20–25 depending on diameter	1,800	30–90 depending on diameter	20–65 depending on diameter
Aluminum alloy 70 bars	4.5	3	1,200–2,700 depending on diameter	330	<10	<10

Table 6.2. *Mass and power loss in tubes*

6.2.2. *Isolating*

It is crucial to be able to isolate components or parts of hydraulic circuits. The first reason for this is that it avoids leakage and fatigue caused by the presence of pressure when components/parts are not being used during a flight phase. For example, this concerns landing gear and high-lift devices when cruising. The second reason is that, in the event of a malfunction occurring, isolation is needed to contain the spread of faults and prevent them from evolving into failures. For example, in the event, an engine fire breaks out or a main pump loses power.

In certain cases, the isolation function simply has to open or close the connection between two hydraulic ports, as depicted in Figure 6.1(a). It therefore behaves similarly to an electric switch. In order to achieve this, a two-port and two-state shut-off valve is used. For instance, this function is performed in pumps:

– on the suction side, to isolate the pump from the reservoir in the event of a fire (fire shut-off valve);

– to isolate the pump from the high-pressure line: in this way, maintenance operations can be performed if needed and the pump flow delivery can be stopped by a blocking valve in the event of a fault. Concerning pressure-compensated variable displacement pumps, the blocking valve is integrated into the pump and the isolation control signal also derates the pressure regulator in order to reduce the required driving torque. This facilitates driving engine start-up [EAT 00].

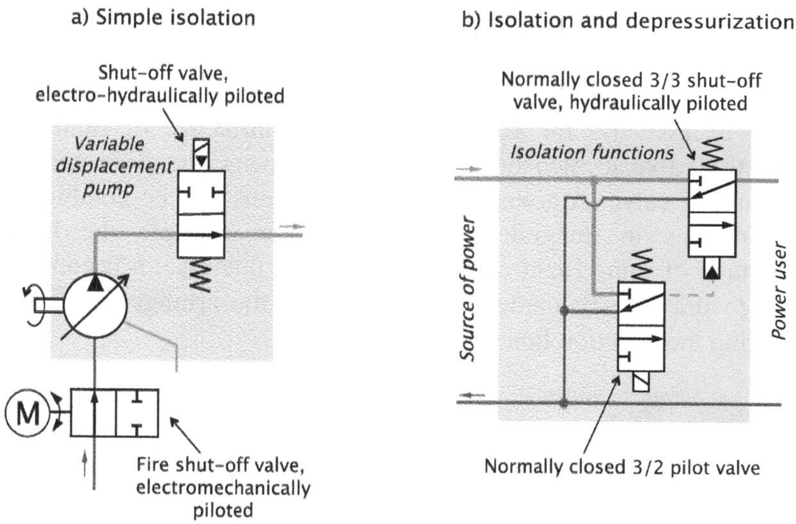

Figure 6.1. *Concepts of isolation from supply and depressurization*

More generally, the isolation function must both cut off the connection with the high-pressure source and drop pressure in the isolated part. It is therefore necessary to use a two-way and two-state (3/2) selector valve in order to connect the isolated part either to the pressure source or the reservoir return line. When valves are required to let a large flow through, they call for large drive forces that are incompatible with a direct electromagnetic pilot. A smaller 3/2 valve is used to pilot the main valve (see Figure 6.1(b)). It plays the same role, as a power relay does for electricity. This configuration is, for example, implemented to isolate slat actuation (Power Control Unit (PCU)) on Airbus A380 [BOW 04].

There also exist purely hydromechanical arrangements which make it possible to automatically isolate the user from the power source. In the event of supply pressure loss, this function contributes to retaining fluid in the actuator in order for the latter to keep assuming secondary functions (damping, anti-cavitation, etc.). Figure 6.2(a) gives an example widely used for flight control actuators, which combines two 2/2 shut-off valves internally piloted by pressure. The connection between the user and the source is conditioned to only open in the presence of pressure in the supply line. This makes it possible to hydraulically pilot the valve located on the pressure line. This valve, in turn, mechanically pilots the valve associated with the return line. When the user is isolated, other components, that are not mentioned on the schematic, maintain a sufficient amount of pressure in order to avoid desorption and cavitation. In the event that the pressure increases excessively, for example, under the influence of fluid thermal dilation, the valve located on the return line opens, therefore acting as a pressure relief valve. Automatic isolation functions can be combined with other functions in integrated components, for example the hydraulic accumulators of some flight control actuators. In this case, it becomes more difficult to draw a schematic representation of the arrangement explicitly highlighting the isolation function.

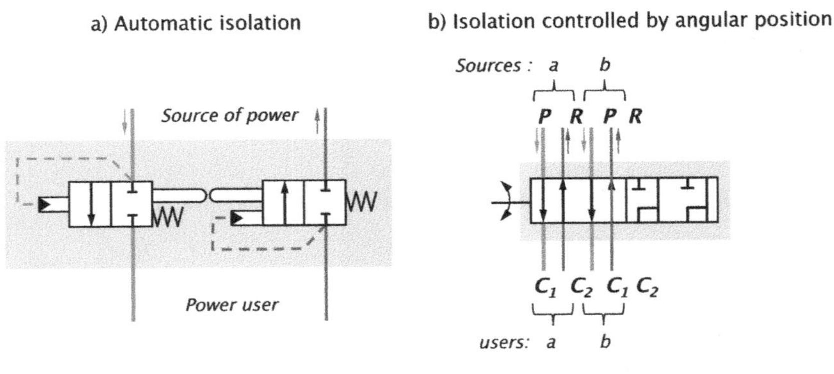

Figure 6.2. *Concepts of isolation from supply and return lines*

The isolation function can also be piloted by the relative motion of two bodies, as shown in Figure 6.2(b). This solution is reliable to prevent inadvertent activation of actuators. For instance, it can be found in nose landing gear, where a rotary isolation valve is mechanically piloted, following a specific kinematic. This configuration makes it possible to allow gear steering and down position locking functions only once the gear is fully extended. Figure 6.2(b) is drawn in the gear-down configuration. In the gear-up configuration, pressure supply is stopped and hydraulic ports of the two users a and b are each connected to the return lines of their hydraulic circuit.

6.2.3. *Sequencing user power supplies*

The hydraulic sequencing function makes it possible to successively power users when they have to perform actuation functions one after the other. In this way, control functions are carried out strictly from the hydromechanical domain. This is especially needed for extending and retracting landing gear. Indeed, it is useful for sequencing the different functions of locking/unlocking, opening/closing of doors and extending/ retracting gear. A simple concept involves on/off valves mechanically piloted by the arrival of the previous actuator at the end-stop. However, this concept is restrictive in terms of geometrical integration because it relies on the need to mechanically sense the actuator or the load position. It is possible to bypass this issue with the help of a sequence valve. Indeed, with such valves, end-stop detection is replaced by pressure level detection in the supply chamber of the previous actuator. Such detection is done based on pressure drop generated by the actuator metering function: this pressure drop only exists in the presence of flow. When the end-stop is reached, the flow becomes zero and full pressure develops in the supply chamber. The sequence valve then opens to enable power supply of the next actuator in line. Despite the advantages it offers, this solution is hardly ever used in aeronautics because it does not absolutely guarantee that the sequencing will be correct.

6.2.4. *Merging sources*

When several sources have the possibility to power one user, an "OR" function has to be established between the two sources (see Figure 6.3). However, the interactions between the two sources have to be limited, and more importantly, sources must be prevented from dispensing power into the other. In order to achieve this on a stand-alone basis, a concept consists of associating two check valves, as depicted in Figure 6.3(a). Check valves act similarly to diodes in electronics. Generally speaking, these valves are calibrated to open and close at a pressure of a few tenths of a bar.

It is possible to establish an order of priority between the sources (see Figure 6.3(b)) if the two check valves have different opening pressures. For example, this is the case on the Rockwell B-1B bomber [AUS 83]. Furthermore, if the two sources 1 and 2 deliver the same pressure, then source 1 has priority over source 2. Then, if source 1 is failing, source 2 automatically takes over. The two sources can eventually operate simultaneously if the operation point allows it.

For the purpose of associating a normal source with a backup source, a shuttle valve that only comprises a single moving part is often used (see Figure 6.3(c)). In normal mode, the normal source feeds the user and the backup source is isolated. The latter replaces the main source if the main source pressure is lower than the backup source pressure. This setup is generally designed in order to prevent the two sources from simultaneously feeding the user. For example, this function is used to feed pistons of the single-cavity brakes of Airbus A380 from normal and backup mode servovalves [DEL 04].

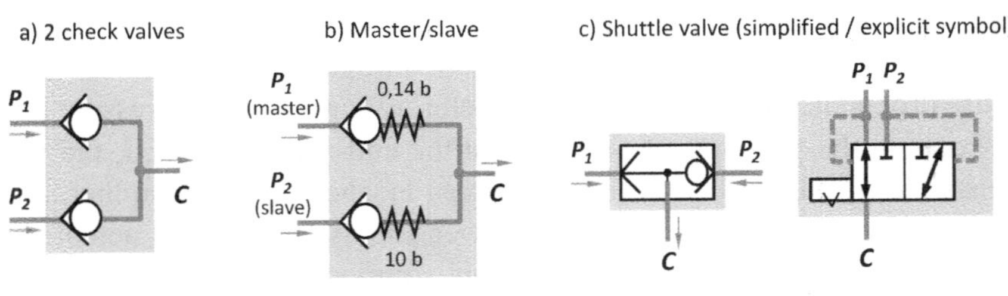

Figure 6.3. *Concepts of source merging*

6.2.5. *Sharing sources*

In the widely used centralized power supply solution kept under permanent pressure and variable flow, users are mounted in parallel (see Chapter 5). Each user is therefore connected to the high-pressure line and draws the flow it needs. This flow-sharing process can follow different configurations: free-flow sharing, sharing with priority or balanced sharing.

Free sharing is simply implemented by connections, as can be seen in Figure 6.4(a).

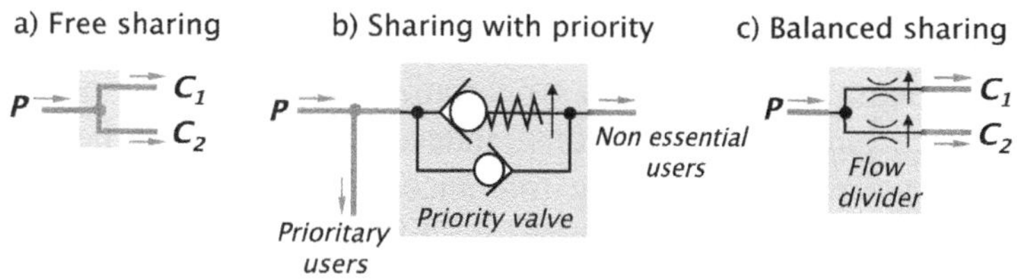

Figure 6.4. *Concepts of flow sharing*

6.2.5.1. *Flow sharing with priority*

Situations may arise where the source is not able to meet the flow demand of users (possible causes include external leakage, too little pump drive speed, etc.). As a consequence, the pump can no longer maintain working pressure and its pressure quickly drops as the flow demand grows. In order for vital users (flight controls for instance) to remain operational in this situation, they are given power supply priority over non-essential users; it is reasonable to do so because the latter assumes non-critical functions or functions that can be completed by backup means (landing gear extension, flap and slat actuation, etc.). Therefore, priority valves are introduced, as shown in Figure 6.4(b). In most cases, priority valves are in fact isolation 2/2 valves (i.e. on/off and shut-off) piloted by their upstream pressure, which exhibits strong hysteresis in order to prevent opening–closing pumping. They have to be mounted in series on the non-priority hydraulic line. In an open configuration, they ought to generate minor pressure drops in order to limit energy loss. For instance, for a supply pressure of 207 bars, the priority valve is fully opened for an upstream pressure greater than 150 bars, completely closed if this pressure drops below 130 bars and starts opening when it exceeds 140 bars. For strong flows, the priority valve is equipped with a pilot valve that facilitates the achievement of good performances. This pilot valve acts by separating pressure threshold detection and power shut-off functions. Furthermore, there are also three-port priority valves that include the essential user port of Figure 6.4(b).

6.2.5.2. *Balanced flow sharing*

Conversely, in other cases, it can become necessary to evenly share flow between users in order to hydraulically synchronize their movements. Very often, simply fitting hydraulic restrictions on every user line does not sufficiently minimize their interaction. Therefore, flow dividers are introduced, as shown in Figure 6.4(c). Their purpose is to compensate for the effects that pressure and temperature have on the flow. Nevertheless, in spite of these compensations, hydraulic synchronization is generally not enough to actually synchronize user positions. Indeed, discrepancies between fluid temperature, wear, manufacturing tolerances and forces developed impact internal leakage rates which are therefore different for each user. Given that position is the integral of velocity, which is itself proportional to the operating flow rate (i.e. the flow supplied by the divider minus the leakage rate), large position discrepancies can quickly appear between users. This is the reason why position synchronization is rarely based on flow synchronization. Regarding secondary flight controls – flaps and slats – a centralized actuator (PCU) associated with a gears and nut–screws mechanical transmission is generally used. Its purpose is to transmit movement to every control surface of the left and right wings. Then, the only possible cause of position desynchronization originates from the mechanical transmission deformation resulting from the influence of transmitted loads. As for hydraulically actuated thrust reversers, such as the ones on Airbus A340–600, the synchronization issue is different. Mounting three cylinders in parallel on each translating cowl is not enough to avoid the seizure of nacelle slide tracks under arching effect. In order to avoid this risk, synchronization is facilitated by interconnecting cylinders with the help of an elastic mechanical transmission (i.e. a flexshaft) that transfers force between actuators, proportionally to their position difference.

6.2.6. *Storing/restoring energy*

6.2.6.1. *Needs*

In a hydraulic system, numerous needs call for the ability to store and restore hydrostatic energy:

– In order to satisfy transient flow demands: In this case, the response delay of pump displacement adjustments has to be compensated for when large transient flows are demanded. Indeed, during certain phases, users can suddenly demand substantially large flows. Although pumps have the ability to supply this flow in steady state, they cannot react instantaneously to the transient demand of a user and there are several reasons why. First, the response time of a pressure-controlled pump to a sudden change in flow is typically of t_r = 100–200 ms. Next, the pressure compensator integrated in the pump only measures the pump output pressure. Therefore, it has no knowledge of the supply pressure at the level of the user which can be located several dozens of meters away. This has two consequences. The first is tied to the propagation velocity of a pressure wave, which is of the order of c =1,000 m/s (see section 3.4.4). Hence, the sudden valve opening of a user located l = 20 m away from the pump will trigger a sudden flow demand. Pressure at the user will drop, and the wave generated will take Δt = 20 ms to reach the pump. The second consequence is due to hydraulic inertia, which has already been introduced (see the example of section 6.2.1.2.2).

– In order to provide a backup energy source: here, the goal is to retain a hydraulic energy source when pumps are inactive, no matter whether this situation is normal or it results from a fault. Maintaining a backup power source is most important for brakes. It is generally asked that, in the event of loss of normal hydraulic power, local hydraulic energy storage can enable a given numbers of successive complete braking actions (five or six for example). In this case, the accumulator quickly restores the pressurized fluid volume required for the preloading phase. This phase starts with the approach of braking pistons to the disk pack and ends with their contact. For instance, on a large passenger jet, this volume equals V = 80 cm^3 per wheel. Next, the accumulator must also supply the volume required to modulate braking forces. In this phase, it is necessary to compensate for both hydraulic fluid compressibility and piston displacements due to the deformation of the brake pack under the influence of the force applied. For example, under full braking force, this deformation represents a volume V = 30 cm^3 per wheel on a large passenger jet.

– In order to maintain pressure at a standstill or after isolation: Isolated circuit parts must be full of fluid at all time and maintained under minimal pressure in order to avoid desorption and cavitation. Maintaining pressure is typically required in two generic situations. First, for parking brakes which have to be provided with a hydraulic pressure source after engines have been shut down in order to keep braking the wheels at parking. The goal of this source is to compensate for possible external leakage and thermal dilation of the fluid and brake disks. Indeed, temperature quickly rises at first because of heat diffusion from the pack to other bodies. It then drops because of convection and radiation with the surroundings, until it reaches ambient temperature. The second situation concerns sub-systems and equipment that are isolated from the rest of the hydraulic system during certain phases of flight (landing gears, slat and flap actuators) or under certain operating modes. The goal is then to maintain the presence of fluid under minimal pressure, therefore avoiding desorption and cavitation. This is also needed for actuators when they operate in passive mode. In this situation, they have to act as dampers for the purpose of preventing *flutter* for flight controls or *shimmy* for nose landing gear steering.

– In order to reduce pulse levels or the magnitude of pressure surges: Here, the aim is to construct a filter that enables reduction of pressure pulses generated by pumps (see Chapter 4) as well as reduction of the water hammer effect (pressure surges) following from sudden flow variations (see section 6.2.1.2.2).

6.2.6.2. *Setup*

In order to satisfy the needs listed above, accumulators (or compensators as appropriate[1]) are introduced. Hydraulic fluid is ill-suited for directly storing hydrostatic energy because it is poorly compressible: when pressure P of a fluid domain drops from 200 to 100 bars, its volume only typically increases by 1%. In order to restore a volume of V = 20 cm^3, an initial volume of V_0 = 2 l would have to be used! Therefore, it is preferential to store elastic energy either in a mechanical spring or pneumatic spring with an inert gas (usually nitrogen). Technology is selected based on the fluid volume that needs to be restored, working temperature range, dynamics, overall dimensions and compression rate (for hydropneumatic solutions). Table 6.3 summarizes the intrinsic strengths and weaknesses of all four types of accumulators. Hydropneumatic accumulators with metal bellows, introduced on Airbus A380, are a real breakthrough because they do not require any maintenance over the entire lifespan of the airplane [DAC 04] (gas cannot diffuse across the metal bellow). This breakthrough has also been implemented with helium instead of nitrogen; this increases the restored volume by up to 44%. As shown in Figure 6.5, accumulator symbols only indicate the implemented energy storage principle (either by mechanical spring or gas compression).

1 The term accumulator tends to be used for storage at high- or mid-range pressures, whereas the term compensator tends to be used for storage at low pressures when the aim is to compensate for volume variations (dilation, external leakage and differential sections) in order to maintain a minimal level of pressure.

Technology	Weaknesses	Strengths
Spring and piston	Weak dynamics due to the moving mass Risk of leakage at the piston Bulky and heavy, except at low pressures	Energy storage independent from temperature Linear behavior Good thermal stability
Pneumatic piston	Weak dynamics due to the moving mass Gas/liquid sealing durability at the piston Highly sensitive to temperature	Compactness
Pneumatic bladder	Pressurization loss by atomic diffusion of the gas across bladder walls Highly sensitive to temperature	Great dynamics
Pneumatic membrane	Limited pressure ratio between extreme operating points Highly sensitive to temperature	Great dynamics
Metal bellow	Bellow balancing Highly sensitive to temperature	No atomic diffusion across metal Long lifespan

Table 6.3. *Intrinsic strengths and weaknesses of different accumulator technologies*

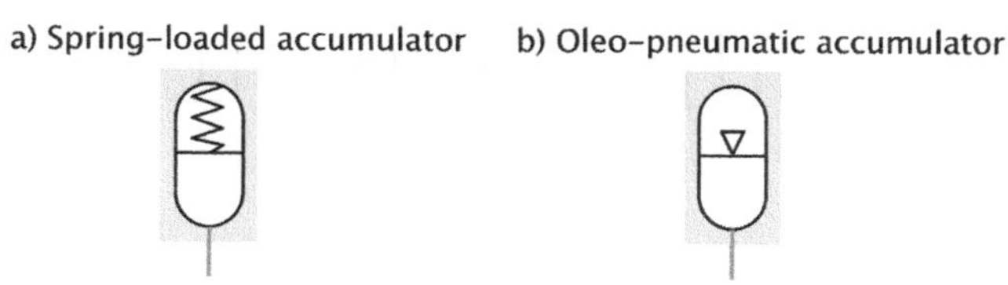

Figure 6.5. *Accumulator symbols*

6.2.7. *Adjusting the pressure level*

In some cases, the pressure level of the hydraulic power distribution network does not correspond to the pressure level required locally. Therefore, pressure has to be increased or lowered in order to adjust the source to the user.

The need to increase pressure mostly concerns ground test benches which are used to perform fatigue and burst tests. For this, hydrostatic transfer units, i.e. *pressure multipliers*, are used. Their operation relies on the Pascal principle by means of different cross-sections for cylinders (see Figure 6.6(a)) and different displacements for rotary machines (motor/pump).

The pressure reduction function is much more frequent. It is constructed by means of a proportional valve mounted in series on the hydraulic line whose opening is piloted by downstream pressure. Using a two-way two-port valve (see Figure 6.6(b)) only enables connection with the pressure source to be metered. Therefore, this valve cannot draw fluid from user lines to prevent its pressure from rising, for instance under the influence of the driving load or thermal dilation. This limitation disappears with a three-port valve (see Figure 6.6(c)), which allows unloading on a low-pressure line. It is important to keep in mind that the pressure reducing valve is intended to evacuate, as pressure loss, all the excess pressure accumulated when supplying the user. Unlike the priority valve, its function is therefore to create a given pressure drop on the path leading to the user. From an energy standpoint, it generates a hydraulic power loss proportional to the supplied flow and to the pressure drop and which is converted into heat.

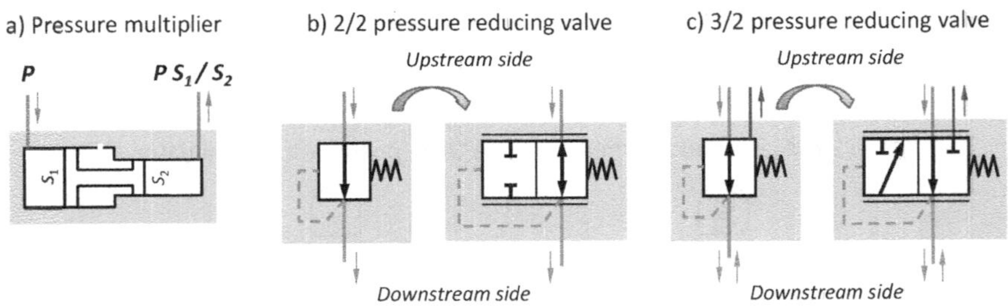

Figure 6.6. *Adjusting the pressure level (simple and explicit symbols)*

6.3. Protecting

Protections are crucial because they assume safety functions. Indeed, they safeguard against consequences on performances and lifespan that follow from abnormal levels of pressure, force, flow and volume.

6.3.1. *Protecting against overpressure/overload*

Protections against overpressure are present in various locations in hydraulic systems. For example, they are responsible for preventing high-pressure lines from getting ruined in the event of a fault in pump displacement adjustment. They are also intended to prevent return lines from getting ruined in the event that low-pressure filters get clogged. At the level of actuators, protection is required in cases where the load becomes driving and forces the actuator to operate in pressure generator mode (for example, this concerns flight controls in the event that a wind burst hits a control surface and auxiliary landing gear steering in the event that a wheel gets impacted by a runway defect). In this case, it is necessary to protect structures and equipment against overload in order to limit the need for mechanical over-sizing. Unfortunately, in practice, the technology to build an inboard force limiting device that is both reliable and light for dozens of kN does not exist. Therefore, the good mechanical efficiency and the low reflected inertia of hydromechanical power transformers (in particular linear cylinders) are put to use. Indeed, they transpose force limitation to the hydraulic domain by means of pressure relief valves.

Pressure relief valves are safety elements. Conceptually, they comprise an infinitely variable position 2/2 valve which is normally closed. It is piloted by upstream pressure or by the upstream/downstream pressure difference. The relief valve has to be mounted in the unloading configuration on the domain that requires protection (see Figure 6.7). When fully opened, a pressure relief valve acts like a fixed orifice.

The static characteristic curve of a pressure relief valve, which is plotted in Figure 6.7(c), highlights several defining quantities: the cracking pressure P_0, the leakage flow before opening Q_0, the fully opened pressure P_n at rated flow Q_n and closing pressure P_1 which defines hysteresis. The calibration of the pressure relief valve is determined by spring preloading. In some cases, it is possible to perform additional functions by tweaking this calibration with the help of an external pilot, either hydraulic or electrical.

In order to protect against overpressures caused by hydraulic fluid dilation in an isolated domain, a specific type of pressure relief valves, so-called thermal relief valves, is used. Those specific valves are sealed in closed position. This is an important characteristic for this application because the fluid volume to let through at opening is very small (typically lower than 10% of the isolated volume). Thermal relief valves can also be slightly dynamic given that fluid temperature varies slowly in general. Their symbol is identical to the symbol of pressure relief valves.

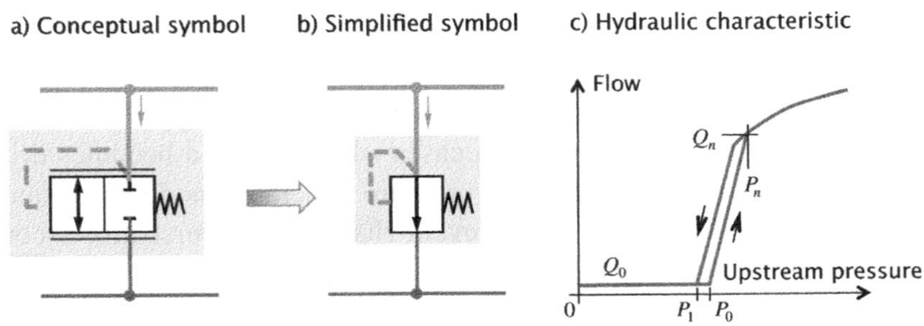

Figure 6.7. *Pressure relief*

Several different arrangements can be implemented in order to perform this protection against overload function. Figure 6.8 introduces four generic solutions for a double-acting actuator. The protection block is inserted in series between the power metering device (for example, a servovalve) and the power transformer (cylinder or motor). Each of these arrangements has different properties:

– Concepts a and b only put to use a single pressure relief valve, but they rely on two check valves. Protection provided is identical for both force (or differential pressure) directions. Solutions c and d use two pressure relief valves, and they have no need for check valves. They allow different protection thresholds if needed.

– Concepts a and c protect in terms of differential pressure between the two power transformer lines. Hence, for a symmetrical hydrostatic transformer, they protect in terms of hydrostatic force F_h which is proportional to differential pressure (see equation [4.1]).

$$F_h = S(P_1 - P_2) = S\Delta P \quad [6.4]$$

These two concepts even protect against underpressure, which will be addressed in the next section. Indeed, when the load becomes driving, the hydromechanical transformer acts as a pump. Pressure rises in one chamber while it drops in the other. Therefore, the fluid drawn from the high-pressure line is used to replenish the low-pressure line. Consequently, these concepts are ill-suited for asymmetrical hydromechanical transformers such as single-rod cylinders. Another important consideration should also be taken into account. Effective force F_c at the interface with the mechanical load is not equal to the hydrostatic force F_h. This is because of inertial and friction forces that arise at the level of actuator moving parts. Cylinders have good mechanical efficiency and low rod mass. Therefore, when they are directly connected to the load, the error between these two forces is insignificant. However, this is not the case in the presence of a speed reducer (gear and nut–screw). Indeed, inertial effect and friction of the hydromechanical transformer are reflected at the load, multiplied by the reduction gear ratio (for dry friction) or even its square (for inertia and viscous friction). In order to illustrate this effect, consider a hydraulic motor combined with a nut–screw system of lead p. It is easy to prove that, in this case, the moment of inertia J of the motor rotor acts as an apparent mass M_a at the load such that:

$$M_a = J(2\pi / p)^2 \tag{6.5}$$

Hence, for a pitch p of 2.5 mm and a moment of inertia $J = 5\times10^4$ kg m^2, motor inertia generates the same inertial effect than a mass M_a of more than 3 tons attached to the load.

– Concepts b and d individually protect lines 1 and 2 because they do not act on the differential pressure ΔP. They apparently do not offer protection in terms of drive force. However, they do carry out this function indirectly when the group hydromechanical transformer/metering valve is symmetrical. For example, this is the case for primary flight controls. In this case, the pressure at control ports C_1 et C_2 vary in opposite ways and quasi-symmetrically with respect to the average supply pressure $P_m = (P_P + P_R)/2$. In the case of a symmetrical cylinder of cross-section area S, this gives:

$$P_1 = P_m + \Delta P / 2 = P_m + F_h / 2S \tag{6.6}$$

$$P_2 = P_m - \Delta P / 2 = P_m - F_h / 2S \tag{6.7}$$

The protection against overpressure in lines 1 and 2 therefore indirectly enables hydrostatic overload limitation for a given supply pressure.

Concepts a and b with check valves are hardly ever used. Concept c is very widely used for symmetrical actuators because it also contributes to the protection against underpressures. However, it has to be replaced by concept d for asymmetrical actuators.

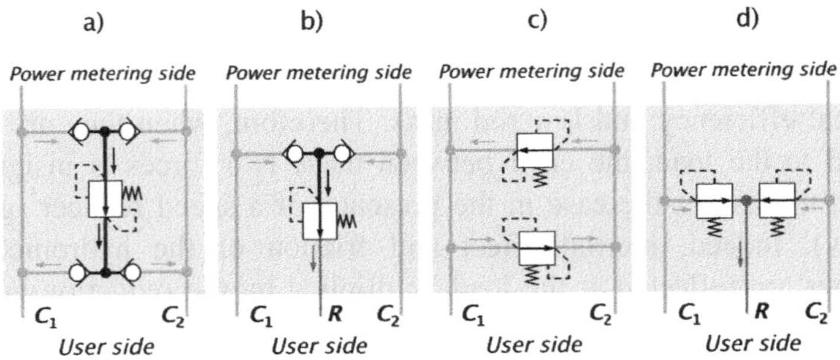

Figure 6.8. *Different protection concepts against overload*

6.3.2. *Protecting against cavitation and desorption*

Gaseous pollution of the hydraulic fluid can be avoided by maintaining a pressure level high enough to repel the risk of desorption and cavitation. Overall pressurization of the hydraulic system through its low-pressure line, if it exists, is performed by the reservoir, as explained in Chapter 3. This solution is generally not sufficient at the level of users, and it requires a charging function for several reasons.

First, certain parts can be isolated from the rest of the circuit because they are not active. This happens when the function performed is not required for the concerned flight phase (for example, landing gear when cruising) or simply when the aircraft is not in use. Therefore, it is necessary to be able to tolerate low temperatures for which the hydraulic fluid retracts and takes up less space. Inactivity can also result from an actuator operating in passive mode. Indeed, under this mode, the actuator lets itself be driven by other systems while acting like a damper. Therefore, it is necessary to maintain a pressure level sufficient for saturation and vapor pressures not to be reached by venturi effect in damping restrictions.

Second, most of the time when users are active, they drive loads that can become driving under the influence of external forces or due to their inertia to direction of movement reversal, for example.

Charging consists of forcing fluid replenishment using a low-pressure source:

– If the user is not isolated from the circuit, the return line naturally acts as this low-pressure source when the reservoir is pressurized. The connections between the low-pressure line and every line that requires charging are opened automatically (see Figure 6.9(a)). This is possible, thanks to a check valve acting as a re-feeding or an anti-cavitation valve. These valves are set to open at a very low differential pressure (a few tenths of a bar for example).

– If the user comprises a symmetrical hydromechanical transformer, it is possible to take advantage of the protection function against overpressure (see Figure 6.9(b)) to partially assume the charging function, as mentioned in the previous section.

– If the user can be isolated from the rest of the circuit, a compensator is introduced, as in Figure 6.9(c). It fills up when the user is active. Then, an isolation function must be added on the return line *R*.

– Certain configurations, such as the one depicted in Figure 6.9(d), combine together compensation and the isolation functions. This enables automatic compensator replenishment. Here, the return line *R*', internal to the actuator, is separated from the metering device supply return line *R*. Following a specific design, it is possible to integrate the 2/2 valve in the spring-return single-acting cylinder.

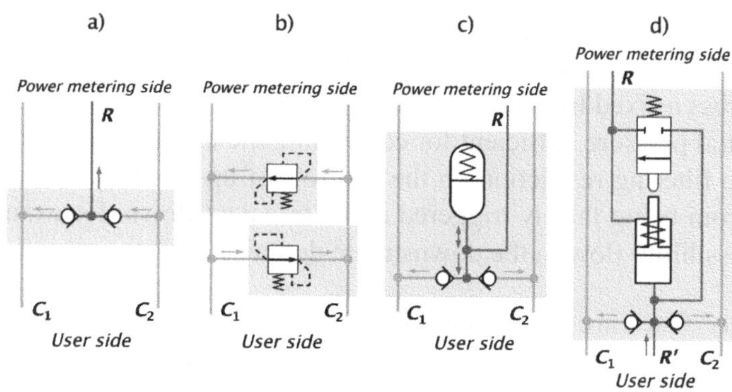

Figure 6.9. *Concepts for preventing desorption and cavitation*

6.3.3. *Protecting against over-consumptions*

Abnormal hydraulic fluid consumption can have catastrophic consequences. This is especially true if it causes a substantial drop in supply pressure (pump inability to supply the demanded flow) or, even worse, if it causes an external leak that empties out the hydraulic system. When this risk is serious, safety measures call for the implementation of protection functions against over-consumption.

The protection function is mounted in series on the line that requires protection. It allows free flow under normal circumstances. In the event of over-consumption in the upstream/downstream direction, the downstream line is isolated and flow is no longer drawn from the upstream line to feed it. This protection is inactive in the reverse flow direction which needs to be free flowing. By analogy to electrical fuses, the component that assumes this protection function is called a hydraulic fuse. A triggered fuse usually remains in this state for as long as the upstream pressure is present. The fuse is then reset when the downstream pressure exceeds the upstream pressure.

The protection function can differ depending on what is needed. It can focus either on the flow consumed or the volume consumed. Generally, flow (or fluid velocity) fuses are used when the line that requires protection delivers a permanent flow in normal mode. This is the case, for example, when it feeds a continuously rotating hydraulic motor. The symbol in Figure 6.10(a) explicitly describes the design concept of flow fuses. Given that it is not possible to directly measure flow in hydromechanics, its effect is measured instead. This is done by simply looking at the pressure drop it causes across a fixed hydraulic restriction. If the flow is too strong, it creates a differential pressure sufficient for controlling the closure of the 2/2 shut-off valve. The filtering restriction on the controls of upstream pressure prevents the fuse from being falsely triggered under transient flows (for example, due to compressibility flow in the downstream domain).

Quantity fuses are often much better suited for protecting control lines whose average flow is zero and whose normal volume consumption is bounded. In other words, they are best for positive displacement (or volumetric) type hydraulic lines with limited stroke. A typical example of this is the protection of brake piston supply lines. Indeed, the risk of rupture is serious for those lines. This is due to the fact that their hydraulic hoses are located in areas prone to projections (landing gear) and are subjected to strong pressure pulses, in particular due to the wheel antiskid system. The required flow during the piston approach phase is both important and functional. On the other hand, the total volume consumed is bounded by the product between the number of pistons, their cross-sectional area and their maximum stroke. The method used to measure the volume consumed is illustrated in Figure 6.10(b). Part of the upstream flow, proportional to the flow crossing the shut-off valve, is drawn in order to feed a pilot piston. Therefore, the piston displacement is representative of the volume transmitted downstream. When it reaches a given position, the pilot piston then sends a control signal to close the valve. As long as they have not been activated, quantity fuses measure the volume supplied to the downstream line. Depending on requirements, resetting conditions can differ and their symbols are represented in Figure 6.10(c). The volume measurement can be reset as soon as the upstream/downstream flow stops (i.e. for a differential pressure of zero). Alternatively, the measure can be saved and resetting only occurs when the flow changes direction (i.e. when differential pressure changes sign).

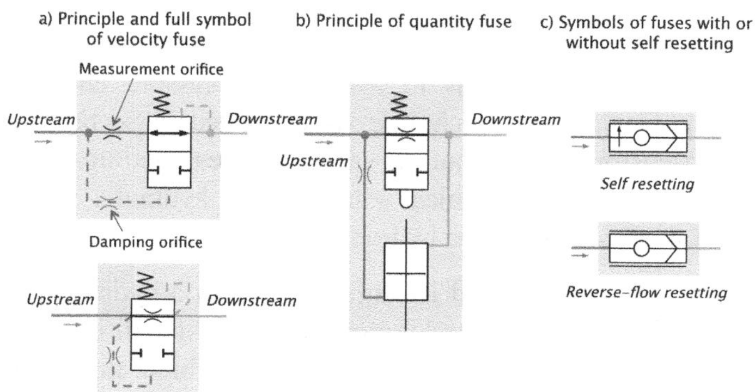

Figure 6.10. *Protection against fluid over-consumption*

Figure 6.11. *Hydraulic fuse (braking system of Boeing B737), (© Chris Brady)*

6.4. Managing the load

Actuators generally have several functional states. They are not only responsible for their primary load actuation function but also for other functions which depend on phases of their lifecycle (phase of flight or maintenance), on their redundancy architecture and on the different faults present. Where appropriate, they must also assume additional functions such as position locking (no load motion), irreversibility (power can only be conveyed from the actuator to the load), load releasing (de-clutching) (no force applied on the load) or damping (actuator develops a force opposite to the load velocity). As with protection against overload, hydraulic technology enables these functions to be performed far more easily, with less weight and with much smaller overall dimensions than a mechanical solution would.

6.4.1. *Locking the load in position*

When performed in the hydraulic domain, locking the load position consists of isolating the hydromechanical transformer from the rest of the network, taking all necessary precautions to avoid underpressure and overpressure. Isolation is implemented on each line, as explained in section 6.2.2. For a double-acting transformer, 2/2 shut-off valves can be paired up. Once again the benefits of replacing the mechanical solution by a hydraulic solution heavily depend on the shortcomings of the hydromechanical transformer:

– if the transformer has fluid leakage, such as in a hydraulic motor, then the external force applied by the load on the actuator will be converted into pressure. This pressure will lead to a leakage rate which, in turn, will enable load displacement. In order to illustrate this effect, consider the example of a trim horizontal stabilizer cylinder actuator using a hydraulic motor with a displacement of 10 cm^3/rev, a rated speed of 4,000 rpm and a leakage rate of 0.4 l/min at the required operating point. The motor must control the pitch of the horizontal stabilizer over a range of 18° with a speed of 0.4°/s, via a mechanical transmission of ratio 1/48,000. Under these circumstances, leakage represents 1% of the rated flow of the motor. Furthermore, if locking is carried out by simple hydraulic isolation, leakage causes a pitch drift of 3° in 10 min, in other words more than 15% of the overall stroke;

– apparent compressibility of the fluid entrapped in the power transformer generates an elastic return effect on the load, in the absence of leakage. As a consequence, the load can travel almost proportionally to the external force.

This effect must be assessed, but most of the time it is negligible. The fact that it gets coupled with inertial effects of the load and the actuator is a much more serious issue. Indeed, such a coupling generates a hydromechanical natural mode. Consequently, in addition to isolation, the load has to be damped.

In the end, hydraulic locking is only implemented when it is enough to satisfy load position locking requirements. If this is not the case, there is no other option but to resort to a mechanical brake.

6.4.2. *Ensuring irreversibility*

The irreversibility function is responsible for holding the load in position in the presence of external forces and when the actuator is not functional. It is implemented in the hydraulic domain by piloted check valves that are arranged according to the layout shown in Figure 6.12 in order to form a hydraulic irreverser. The irreverser is inserted in series between the metering valve and the hydromechanical transformer. It is placed as close as possible to the latter in order to minimize the possible influence of leakage from other components. If the load is driving, the hydromechanical transformer acts as a pump. In the absence of pressure on the metering side, the line it pressurizes is then closed by the check valve. Conversely, when a differential pressure is created on the source side between lines 1 and 2, the line under pressure forces the opposite valve to open, allowing return of the transformer fluid. Therefore, the irreversibility function assumes the isolation function when the load is driving. Similarly to the isolation function, it suffers from the same limitations in terms of leakage and compressibility. Consequently, a mechanical irreverser (or no back) should be added between the actuator and the load if back drivability constitutes a catastrophic failure.

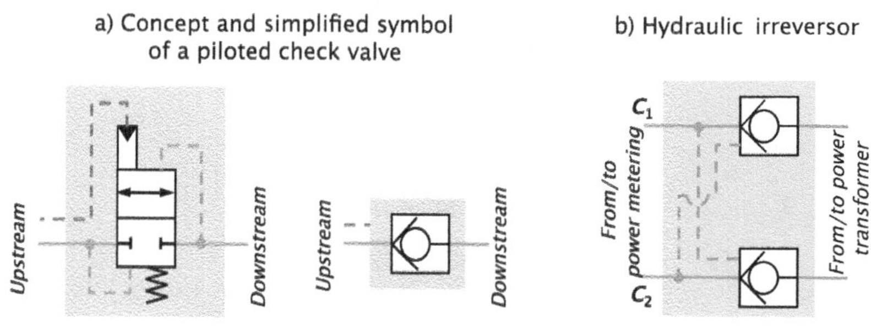

Figure 6.12. *Concepts for ensuring irreversibility*

6.4.3. *Releasing the load*

When the actuation function is not active, it can be necessary to de-clutch (or release the load), meaning having no force applied on the load by the actuator, which in turn means the actuator can be dragged along by the load. This is called backdrivability. This function is often required for maintenance operations. It is also found in nose landing gear steering actuators, in order to allow rolling by pushback or towing at parking. Hydromechanical de-clutching consists in cancelling hydrostatic forces generated by the hydromechanical transformer when the load is driving. Its implementation differs depending on whether it is symmetrical or not. Figure 6.13(a) shows the implementation of the de-clutching function in hydraulic form, in the case of a symmetrical user. Load release is pressure controlled, for example it can be activated by the supply pressure. De-clutching then consists of connecting the chambers of the hydromechanical transformer with the help of a 2/2 shut-off valve in order to bypass them. Figure 6.13(b) uses the same hydraulic piloting but for an asymmetrical user. In this case, chambers are both connected to a low-pressure source, for example the return line. In de-clutched mode, it is necessary to make sure that the metering function has stopped feeding the user. Then, it is possible to combine together an isolation function and a de-clutching function in a single four-way two-position selector valve, as shown in Figure 6.13(c). This example is representative of the implementation of the function for primary flight control electro-hydraulic actuators. De-clutching in hydraulic form is activated by the electrical power supply of a solenoid valve which hydraulically pilots the valve. Orifices on the metering side are closed when de-clutched mode is active. In the case of electro-hydrostatic actuators, they can also bypass the two ports on the source side.

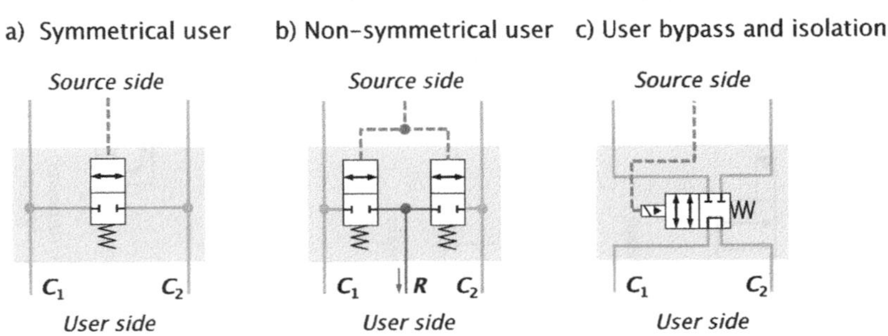

Figure 6.13. *Concepts to release the load (also called load de-clutching)*

Performing de-clutching in the hydraulic domain introduces parasitic resisting forces. They first originate from friction in mechanical transmission toward the load and in hydromechanical transformer dynamic sealing. Second, they result from pressure drops generated by bypass flows in lines and valves. Those losses reflect at the load, magnified by the overall transmission ratio (motor displacement or piston cross-section added to the mechanical reduction ratio). However, these parasitic forces contribute to the load damping function if it is required beyond a simple de-clutching.

6.4.4. *Damping the load*

The load damping function consists of having the actuator generate a dissipative force in response to load movement solicitation. This damping function is performed differently depending on circumstances: when the actuator is active and in the presence of a driving load, in release mode when the load has been hydraulically de-clutched, to avoid load vibrations or to soften shocks with mechanical end-stops.

6.4.4.1. *Damping a driving load when the actuator is active*

When the actuator is placed in a position servo-loop, driving load damping is performed actively by the control signal (if its bandwidth allows it). However, if the actuator is active but has on/off controls, (for example for landing gear extension), damping is carried out by inserting a restriction or a flow control valve on the return line of the hydromechanical transformer.

6.4.4.2. *Damping a hydraulically released load*

In order to damp the hydraulically released load according to concepts described in section 6.4.3, it is simply a matter of replacing the free path by a restricted path. This can be integrated at the de-clutch valve. Given that the flow in the restriction quickly becomes turbulent, the damping force F is proportional, of coefficient f_q, to the square of the relative speed v of the load with respect to the actuator body. This result can also easily be deduced by combining the transformation equation for a linear cylinder (equation [4.1]):

$$Q = Sv \qquad [6.8]$$

$$F = S\Delta P \qquad [6.9]$$

with the damping restriction equation (equation [5.14]):

$$Q = k_o \sqrt{|\Delta P|} \text{sign}(|\Delta P|) \text{ avec } k_o = C_{q\infty} s \sqrt{2/\rho} \qquad [6.10]$$

This finally gives:

$$F = \frac{S^3}{k_o^2} v^2 \text{sign}(v) = f_q v^2 \text{sign}(v) \qquad [6.11]$$

The amount of energy dissipated over one load sinusoidal oscillation cycle of magnitude x_0 and angular frequency ω (so of frequency $f = 2\pi/\omega$) is therefore equal to:

$$E = \frac{8}{3} f_q x_M^3 \omega^2 \qquad [6.12]$$

Consider a short cylindrical damping orifice of diameter d = 1.5 mm, of limit flow coefficient $C_{q\infty}$ = 0.75 (in established turbulent flow regime) associated to a cylinder of hydrostatic cross-sectional area S = 20 cm^2 running on a fluid of specific density ρ = 1,000 kg/m^3. It generates a quadratic damping factor f_q = 0.228 daN/(mm^2/s^2). Therefore, at a frequency of f = 2 Hz and for a range of motion of x_M = 5 mm, the energy dissipated by the restriction over one oscillation cycle is equal to E =120 J and the maximum differential pressure during the oscillation is equal to ΔP_M = 45 bar.

An alternate solution consists of taking advantage of the presence of pressure relief valves on control lines 1 and 2. Indeed, when the load has been released and the actuator is not active, it is only a matter of derating relief valves to make them open at a lower differential pressure ΔP_0. Then, the damping force is obtained from the differential pressure:

$$\Delta P = \Delta P_0 \text{sign}(v) \qquad [6.13]$$

and equals:

$$F = S \Delta P_0 \text{sign}(v) = F_C \text{sign}(v) \qquad [6.14]$$

Therefore, in this situation, the damping force is similar to a dry friction force F_C, and the amount of energy dissipated over a cycle does not depend on frequency. Hence, it can be expressed as:

$$E = 4x_M F_C \tag{6.15}$$

Consider the same conditions as before and consider a derating $\Delta P_0 = \Delta P_M$ corresponding to the previous maximum pressure equal to 45 bars. This time, the energy dissipated over one oscillation cycle is E =180 J.

It is worth noting that if the damping forces were purely viscous, meaning proportional of coefficient f_l to speed v, then the amount of energy dissipated on a cycle would be given by:

$$E = \pi f_l x_M^2 \omega \tag{6.16}$$

For the same maximum pressure as in the above example, it would equal E = 141 J.

These two solutions are, for example, put to use on the flight control actuators of Airbus A330-200: the first is implemented for ailerons and the second for the rudder.

6.4.4.3. *Avoiding load vibrations*

Nose landing gear steering is subject to shimmy vibrations, the consequences of which can be catastrophic. In order to damp these vibrations, it is necessary to dissipate some energy. This function is performed by a shimmy damper, active at all times no matter the circumstances, which is mounted in series on each of the control lines. As can be seen in Figure 6.14, it consists of three elements in parallel: a fixed restriction, a check valve which bypasses the fixed restriction for flows entering the hydraulic transformer and finally a pressure relief valve which also contributes to dissipating energy. Under negative flows, when the hydromechanical transformer chamber fills up, there is low resistance to flow. In fact, this resistance corresponds to the one created by the open check calve that, in this situation, acts like a fixed restriction with a large cross-sectional area. Under positive flows, when the chamber empties out, the fixed restriction generates turbulent pressure loss up to point A where the pressure relief valve starts opening. Starting from point B, the relief valve is fully open and the unit formed by restriction + relief valve acts a fixed restriction (flow depends on the square root of the differential pressure, in turbulent flow regime).

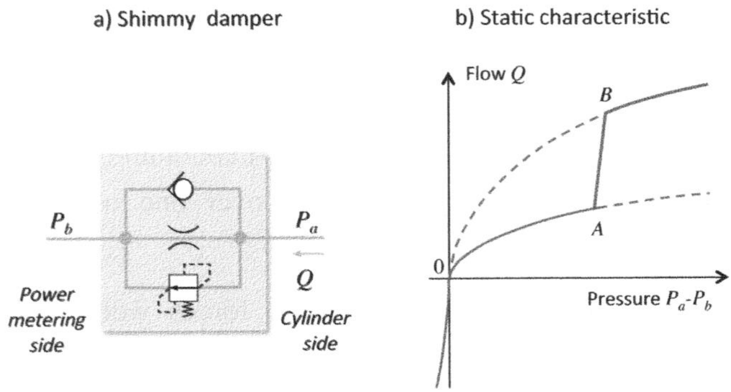

Figure 6.14. *Shimmy damper*

6.4.4.4. *End-stop snubbing*

In on/off position actuation functions, the load travels until it reaches a mechanical end-stop. For integration reasons, it can be preferred for the end-stop to be placed inside the actuator itself. When reaching the end-stop at a relative speed v, the kinetic energy of the moving part of mass M has to decrease until it reaches zero, which corresponds to the load being at a standstill. In the absence of particular dispositions, this energy gets converted into elastic energy by the deformation of parts in contact that have mechanical stiffness K. The peak force F is therefore given by:

$$F = v\sqrt{KM} \qquad [6.17]$$

Based on the values of contact stiffness in practice, the peak force is very often inadmissible. In order to reduce kinetic energy upon arrival at the end-stop down to a few Joules, an end-stop snubbing function is implemented. It gets activated only when approaching the end-stop. The design concept used to implement this function is depicted in Figure 6.15. During the phase of approach, proximity of the end-stop is detected by the relative position of the hydromechanical transformer moving part. The detection validates activation of a hydraulic restriction, which is sized to cause the intended deceleration. Kinetic energy of the load is therefore dissipated as heat by the restriction. During the distancing phase, the restriction is inhibited by a check valve so as not to slow down the load. Depending on implementations, this function is either integrated in the body (extension/retraction for example) or the cylinder rod (thrust reverser for example).

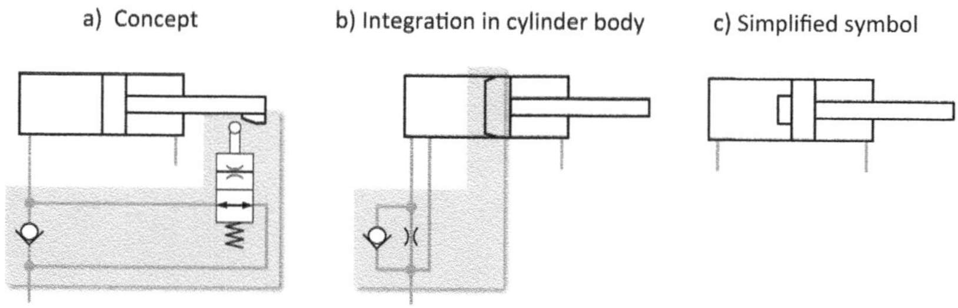

Figure 6.15. *Snubbing at end-stop*

7

Architectures and Geometric Integration of Hydraulically-supplied Actuators

7.1. Introduction

This chapter deals with hydraulically-supplied actuation systems from the perspective of power architecture and geometric integration (or topology). It describes how their elements are combined together and how they are spatially arranged. Key aspects can then be identified as follows:

– arrangement of actuation functions throughout the aircraft;

– architecture and routing of hydraulic power networks;

– integration of components and equipment; and

– integration of actuators between the airframe and the driven load.

Evidently, the concept of integration extends beyond the geometry aspect. Integration must also satisfy the many requirements and constraints that apply to interactions with the environment and the other systems. For example, this entails:

– resisting aggressions (vibratory, electromagnetic, thermal, etc.) and minimizing emission of such types of aggressions;

– allowing installation, diagnosis and maintenance; and

– complying with segregation, diversity and independence constraints of redundant channels.

7.2. Arrangement of actuation functions

The most critical actuation functions, such as primary and secondary flight controls, are made extensively redundant in order to reach the required level of reliability. Redundancy applies to all elements involved in performing the function, provided they contribute substantially to the risk of failure. Such elements include control surfaces, actuators, their power supply and their controls. The appropriate number and locations of control surfaces are determined, first and foremost, by regulatory aircraft maneuverability requirements depending on the phase of flight (roll rate for example). These are joined by additional constraints relative to the structure, such as doubling control surfaces if they are subjected to exceedingly large deformations or to minimize the risk of aero-elastic coupling. Ailerons are a good illustration of this last constraint. Indeed, wingtip ailerons (outboard ailerons) are very efficient because the leverage of their aerodynamic action to initiate roll is significant. However, low torsional stiffness at the wingtip can attenuate or even reverse the effect of ailerons, when steering ailerons substantially impacts the local pitch of the wing (reversed aileron effect). In that regard, at high speeds, it is best to bring ailerons closer to the fuselage (inboard ailerons), in an area where this effect does not occur. It is now easy to understand why the optimization of the topology selection for control surfaces is a strongly iterative process.

An example of arrangement is given by Figure 7.1 for the primary flight controls of Airbus A330-200. All actuators are hydraulically-powered. Most are exclusively controlled by electrical signals, following a *Fly-by-Wire* type architecture [TRA 06]. Each control surface is symbolized by a rectangle or a square which indicates how many actuators are associated with it, which hydraulic network supplies these actuators (either green, yellow or blue) and which computer controls them (either prim 1–3 or sec 1 and 2). In the event of an electrical outage, rudder and trim horizontal stabilizer (THS) actuators can also be mechanically signaled based on the pilot's setpoints respectively acting on pedals and on pitch trim wheel. Roll control relies on two ailerons per wing, each being driven by two actuators operating in the active/damping mode. Only a single control surface is used for yaw control. However, it is driven by three simultaneously active (active/active/active) actuators that are mechanically signaled by the position output by the yaw damper. The trim horizontal stabilizer is driven by two active/active torque summing hydraulic actuators. The valves associated with these actuators are controlled by three electric motors or by a mechanical signal output by the cockpit. Depending on the phase of flight, spoilers can be used either for roll control, speed-braking or maneuver load alleviation. Therefore, reliability requirements are satisfied by combining three hydraulic power sources, five flight control computers and two mechanical backup channels for transmitting control signals to actuators. Furthermore, the three hydraulic power sources are spatially allocated to the different actuators in such a way as to retain primary flight controls in the event that two of the three sources are lost.

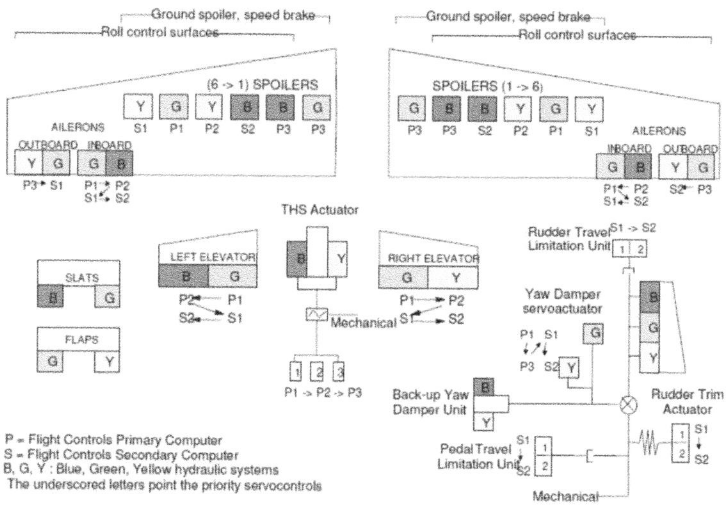

Figure 7.1. *Layout of flight control actuation onboard Airbus A330-200 (© Airbus). For a color version of the figure, see www.iste.co.uk/mare/aerospace1.zip*

7.3. Architecture and routing of hydraulic power networks

7.3.1. *Architecture*

An example of an architecture used for hydraulic power generation onboard Airbus A320 is provided by Figure 7.2. It is a good illustration of the concepts of independence and diversity. Thanks to the hydrostatic reversible power transfer unit, it is possible to exchange power between the green and yellow circuits in both directions without any fluid exchange. This removes the risk of spreading leakage and solid pollution from a circuit to another. Therefore, it is possible to draw a vertical line that completely separates the three hydraulic circuits. This explicitly proves their mutual independence in the hydraulic domain. The diversity of hydraulic power generation functions is satisfied by combining pumps driven either by the engines (engine pump), electrically (electric pump) or, as a last resort, by a ram air turbine. Furthermore, each circuit is equipped with a priority valve so as to prioritize hydraulic power supply of brakes and primary flight controls over landing gear, slats, flaps, and backup electrical power unit. It should be noted that thrust reversers are also hydraulically-supplied.

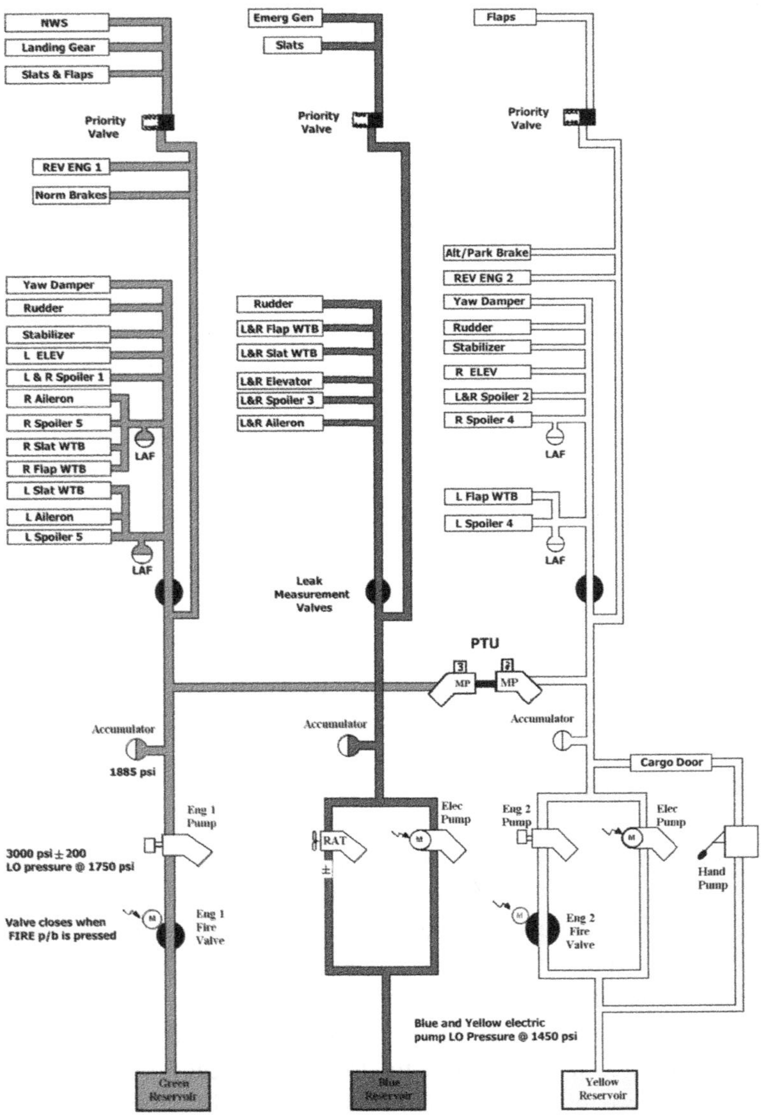

Figure 7.2. *Architecture of hydraulic power generation onboard Airbus A320 (© Airbus)*

7.3.2. *Routing*

Regarding geometrical integration, physical or spatial segregation prevents all channels from being impacted by a single common external cause, like fire or structural damage (bird strike, tire burst, etc.). History has shown the importance of spatially segregating the different hydraulic circuits: in 1981 loss of three of the four hydraulic circuits of a Lockheed L1011 due to the detachment of the main engine fan [BUR 82], in 1985 loss of the four hydraulic circuits of a Boeing B747 due to the rupture of the rear bulkhead [TAK 87], in 1994 loss of the three hydraulic circuits of a Tupolev TU254 due the main engine catching on fire [SEW 94], in 2003 loss of the three hydraulic circuits of an Airbus A300 due to a missile strike [MAL 04]. Figure 7.3 provides two examples of routing ensuring spatial segregation. The first concerns the three hydraulic circuits (red, blue and green) of the Agusta–Westland (formerly Bell–Agusta) convertible AW609. The second concerns the three hydraulic circuits (green, yellow and blue) of Airbus A330.

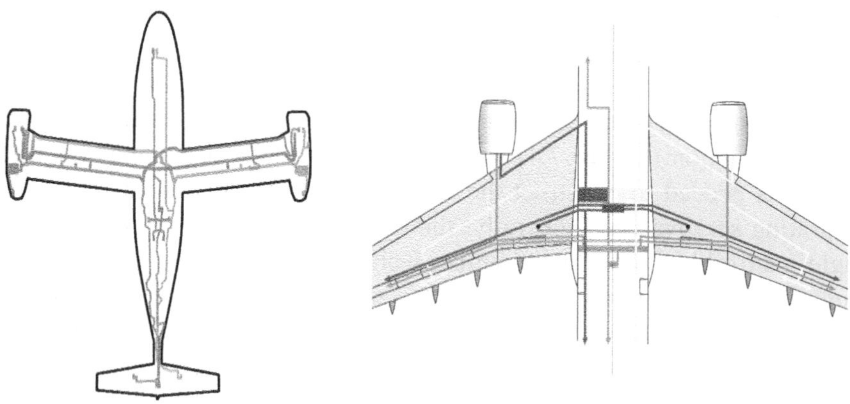

Figure 7.3. *Example of hydraulic circuit routing ensuring spatial segregation Left: AW 609 tilt rotor [FEN 05] Right: Airbus A330 (© Airbus). For a color version of the figure, see www.iste.co.uk/mare/aerospace1.zip*

7.4. Integration of components and equipment

It is possible to identify several different levels and solutions for the integration of hydraulic components and equipment in systems (generation/ distribution, flight controls or landing gears). Starting from the weaker integration levels, there is the in-line integration, the manifold integration and the sub-system integration.

7.4.1. *In-line integration*

In an "in-line" set-up, components are simply inserted on the concerned hydraulic line, as illustrated in Figure 6.11. This solution, although seemingly simple, actually requires a great deal of clamps, hydraulic fittings and tubes. Generally, this goes against reliability, maintainability and compactness.

7.4.2. *Manifold integration*

A manifold is a machined hydraulic block which bundles together hydraulic components in a single spot and interconnects them hydraulically by means of internal channels, thus removing the need for so many clamps, fittings and tubes. The geometrical interface between the block and components is either cylindrical or planar. In the first solution, the manifold sets up radial hydraulic connections with the component: "cartridge" integration requires that the component be inserted axially and that it be secured in place by screws or clamps. An example of this approach is given in Figure 5.31. It concerns the direct drive valve used for nose landing gear steering controls of the Dassault Rafale. The second standardized subplate interface solution (Figure 7.4) reduces the number of costly manufacturing operations needed to construct the manifold. It is the preferred solution for integrating servovalves. Generally, the two solutions are combined within a single manifold in order to minimize weight and overall dimensions of the equipment. The "Viking" type interface provides a single component with both a planar connection for the hydraulic part and a cylindrical connection with an axis perpendicular to the plane for electrical signals. This limits the amount of flying electrical wiring.

However, the manifold must be designed to comply with the manufacturability constraints of channels. Nowadays, these are obtained by machining. However, emerging additive manufacturing techniques [KAU 10] remove many constraints relative to the shape and location of channels. Therefore, when applied to manifolds, this new technology will soon help reduce the weight and manufacturing costs of manifolds.

Manifolds are extensively used to bundle together functions of distribution (isolation valves, etc.), fluid conditioning (filters, etc.), safety (pressure relief valve, check valves, etc.) and monitoring (pressure sensors, etc.). The hydraulic system then consists of an assembly of equipment (pumps, actuators, etc.) and manifolds (high pressure, low pressure, reservoir pressurization, brakes, landing gear, etc.). A generic example is given in Figure 7.5 for hydraulic power generation.

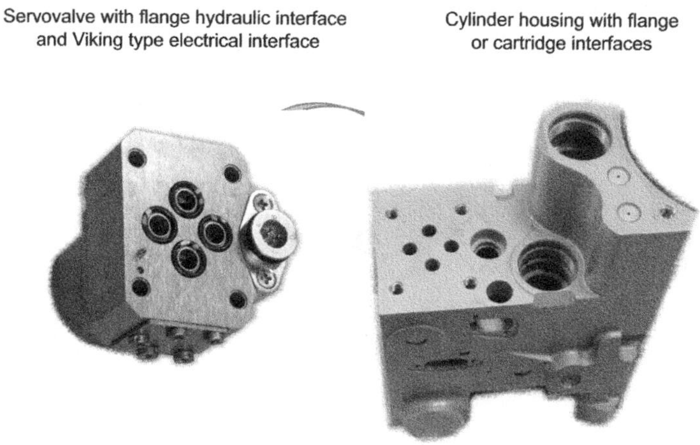

Figure 7.4. *Interfaces for integration of components*

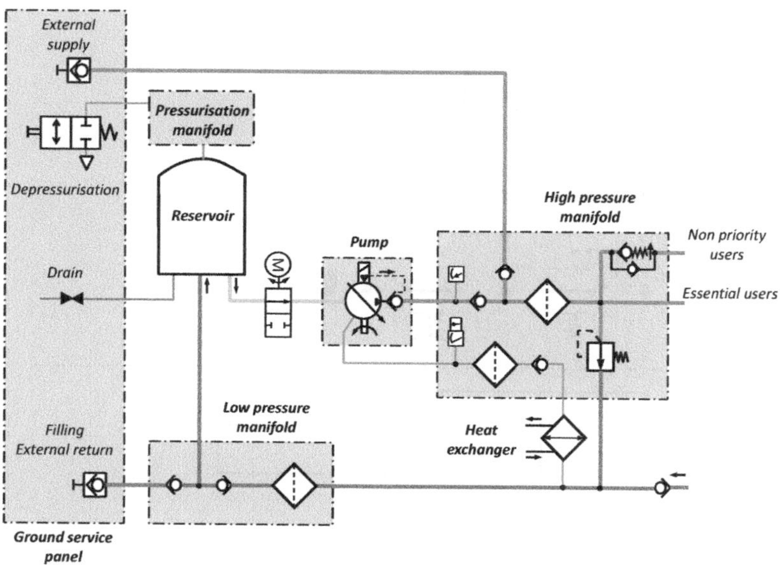

Figure 7.5. *Example of manifold integration for hydraulic power generation and distribution [FAA 12]. For a color version of the figure, see www.iste.co.uk/mare/aerospace1.zip*

Actuator housings may themselves be used as a manifold to integrate the components associated with the various functions required for actuation. Figure 7.6 provides a generic example of manifold integration for a flight control actuator in active-standby mode. The actuator housing acts as a manifold that integrates the following: the conditioning and isolation functions (area 1), the power metering function (area 2), the mode selection function (area 3) and the two protection functions against desorption or cavitation and against overpressure or overload (area 4).

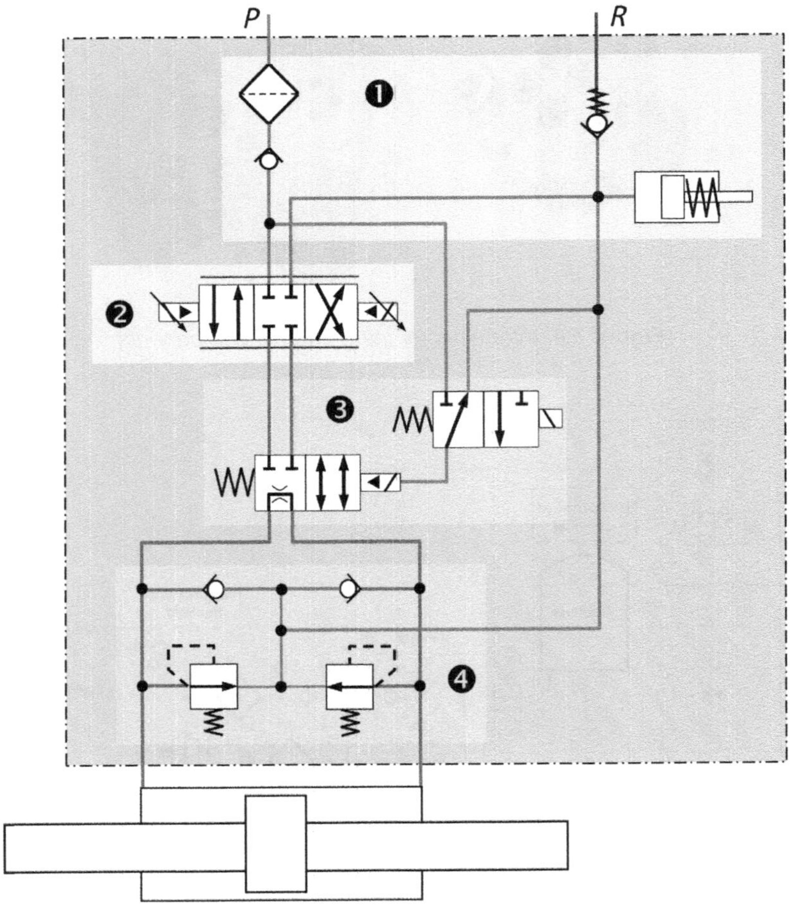

Figure 7.6. *Example of manifold integration for an active-standby actuator. For a color version of the figure, see www.iste.co.uk/mare/aerospace1.zip*

The pictures in Figure 7.7 illustrate this integration for slat and rudder actuators of Airbus A380.

Titanium hydraulic manifold of an Airbus A380 EHA

Titanium hydraulic manifold of Airbus A380 slats PCU, from [BOW 04]

Figure 7.7. *Integration of functions in the actuator housing*

It is also worth mentioning that in certain recent designs of small-size jets, such as the executive Gulfstream G650 jet, sizing constraints require separating the cylinder from its manifold.

7.4.3. *Sub-system integration*

In its most advanced form, integration makes it possible to bundle together all the elements of a system as a single unit. Therefore, all that is left to do for the aircraft manufacturer is to connect this sub-system to its power interfaces (mechanical, hydraulic, electrical and eventually pneumatic) and to its signal interfaces (controls, monitoring). Three recent examples are provided to illustrate this integration approach: the local electro-hydraulic generation system (LEHGS) of Airbus A380 [DEL 04] in Figure 7.8, the main rotor hydraulic subassembly (MHRS) of the Bell 429 helicopter in Figure 7.9, and the aft strut fairing module of Boeing B787 in Figure 7.10.

Figure 7.8. *Local electro-hydraulic generation system (LEHGS) of Airbus A380*

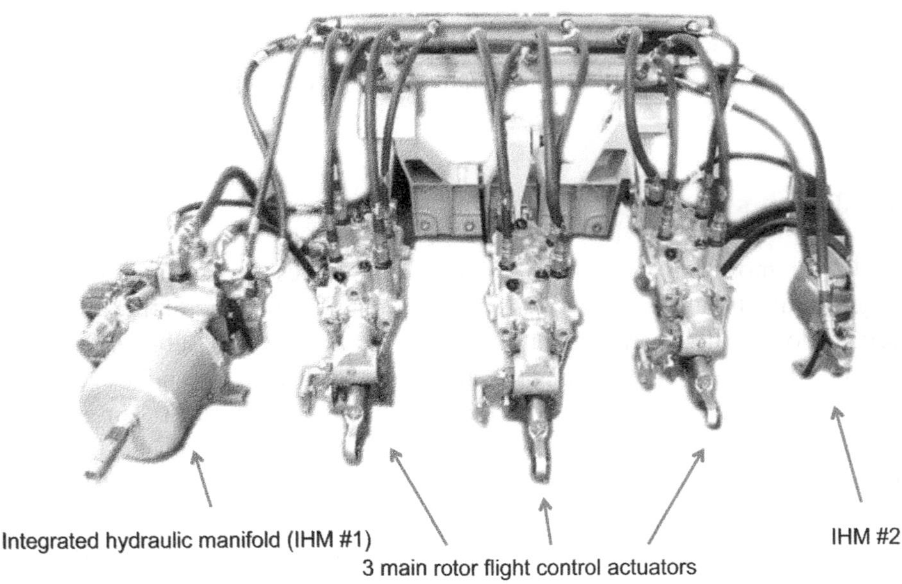

Figure 7.9. *Main rotor hydraulic subassembly (MHRS) of Bell 429*

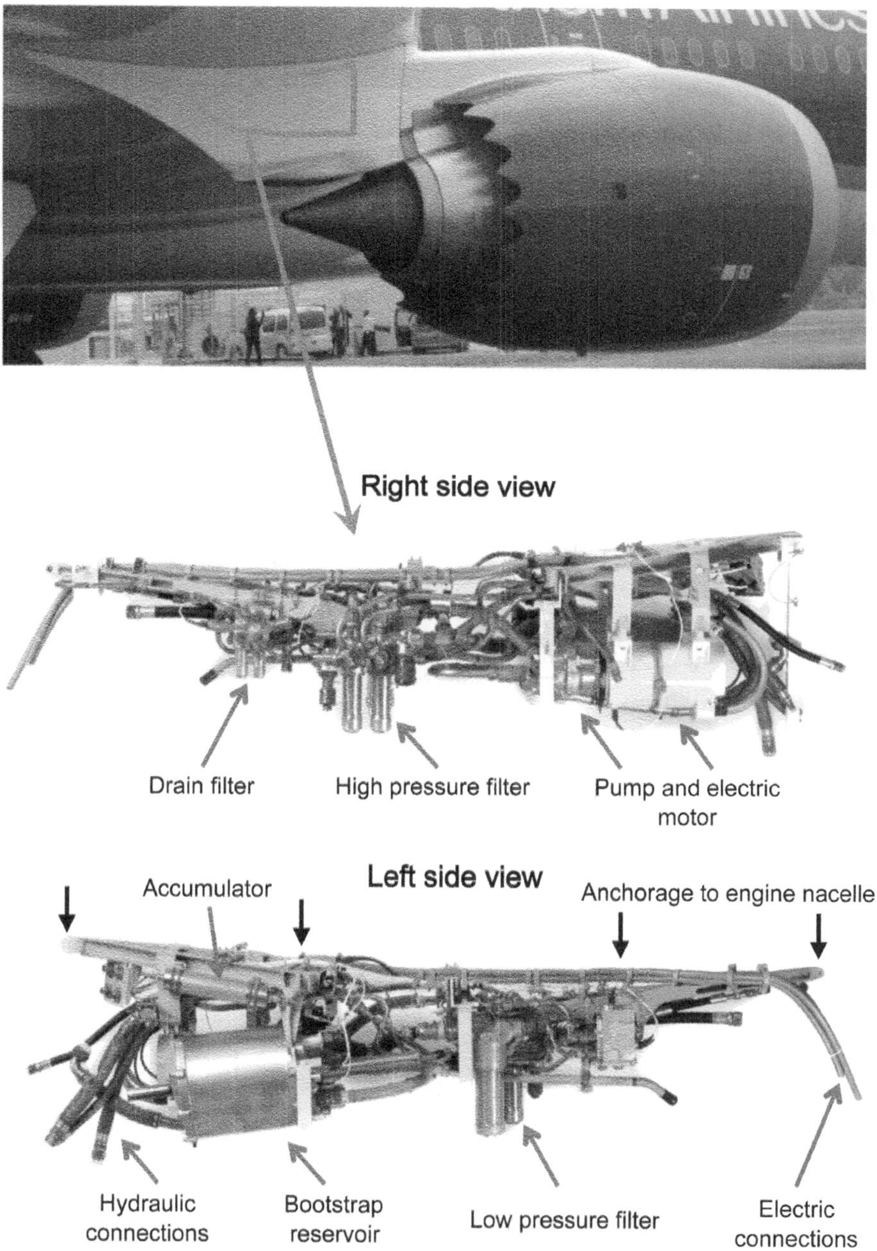

Figure 7.10. *Aft strut fairing module of Boeing B787 (side views © Parker)*

7.5. Integration of actuators in the airframe

7.5.1. *Controls*

For actuators with a mechanical control signal input, geometrical integration must allow the position servo-loop to be established. The aim is therefore to act on the error between demanded position (input displacement) and actual position (position of the actuator moving part that is connected to the driven load). This difference in position is what controls the opening of the hydraulic power valve. Figure 7.11 depicts the two generic geometrical integration concepts for a linear actuator, depending on whether the moving part is the actuator body (moving-body) or the actuator rod (moving-rod).

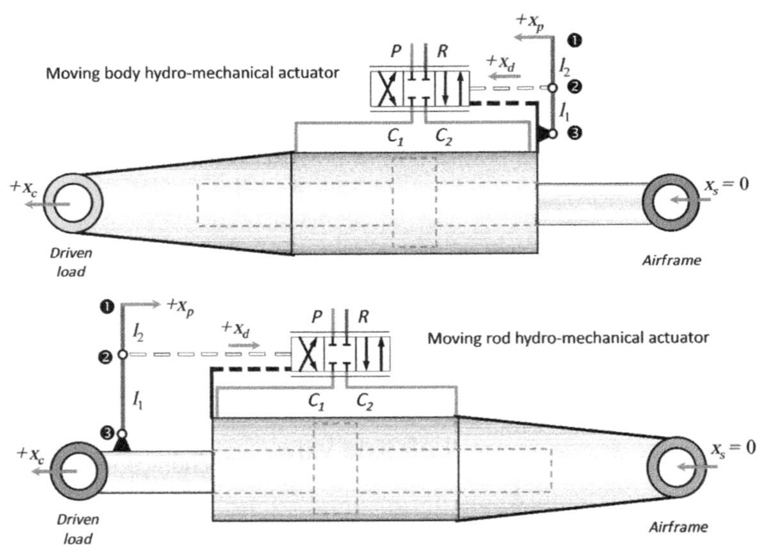

Figure 7.11. *Two concepts of hydro-mechanical actuators*[1]*. For a color version of the figure, see www.iste.co.uk/mare/aerospace1.zip*

7.5.1.1. *Moving-body linear actuator*

In this solution, the position setpoint x_p is applied at point 1 of the control lever. The position of the load x_c is measured through the position of the body at point 3. The cylinder rod, which is connected to the airframe, is functionally motionless ($x_s = 0$). The control lever compares these two

1 Since there is no standard symbol to represent displacement of the valve body, it is represented here by a black bold dotted line connected at the base of the valve symbol.

positions in order to deduce the difference in position. Then it creates a valve opening y (relative spool/body position) proportional to this error:

$$y = x_d - x_c = k(x_p - x_c). \quad [7.1]$$

Kinematic gain $k = l_1/(l_1 + l_2)$ is defined by lengths l_1 and l_2 of the control lever. When the desired position is reached, the hydraulic valve closes and the lever becomes vertical on the figure: the corresponding point 2 therefore travels like setpoint point 1. The load also travels in the same direction as the setpoint. This solution is simple, however there are two disadvantages to this approach:

– the hydraulic connecting ports of the body travel along the actuator stoke, imposing the use of long hydraulic hoses; and

– since it is the actuator body that moves, the moving mass reflected at the load is large. This can limit the dynamics of position controls. Nevertheless, this effect is often acceptable in flight controls of commercial airplanes.

7.5.1.2. *Moving-rod linear actuator*

In the moving-rod solution, the position of the load is now measured through the position of the rod at point 3 of the control lever. The body is connected to the airframe and it is functionally motionless ($x_s = 0$). When the desired position is reached, the hydraulic valve closes. Therefore, the lever is angled in such a way that point 2 is located at the hydraulic zero position of the valve, in other words still in the same spot as on the figure. Hence, when the position difference is zero, point 2 is invariant, no matter the position setpoint. As a result, the position setpoint and the position of the load travel in opposite directions. According to the sign conventions defined in the figure, valve opening $y = x_d + x_s$ can now be expressed as:

$$y = x_d + x_s = \frac{l_1}{l_1 + l_2} x_p - \frac{l_2}{l_1 + l_2} x_c + x_s. \quad [7.2]$$

The presence of the airframe displacement term, x_s, in this equation introduces a "structural" feedback. This feedback limits the impact that the compliance of the airframe has on the stability of the position servo-loop. On the other hand, it also introduces a static error under load [GUI 72]. This moving-rod approach eliminates the two drawbacks of the moving-body solution. First, the range of motion of hydraulic connecting ports is brought down to the small rotations introduced by pin-to-pin mounts described in the next paragraph. Second, the moving mass of the actuator reflected at the load, is now confined to the rod mass instead of the body mass. However, it can be seen that the position of the airframe now directly influences the valve opening.

The integration of the moving-rod concept is facilitated by adopting the alternative implementation depicted in Figure 7.12. However, this comes at the cost of more complex kinematics. The position setpoint is introduced at point 1 by absolute displacement x_p. The position of the rod x_c at point 3 is transmitted to point 4 by the position feedback link. The valve opening y, i.e. the relative spool/sleeve position $y = x_d - x_s$, is implemented by the interaction between the control lever (displacement of point 5) and the structural feedback link (displacement of point 2). In order to meet isostatic conditions, the position feedback link must be ball-jointed to the rod (or be elastic in the transverse direction).

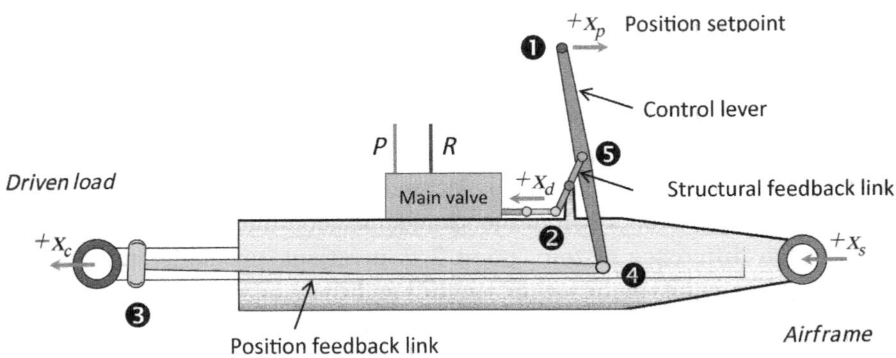

Figure 7.12. *Moving-rod hydro-mechanical actuator*

This solution is widely used for active/active tandem actuators of helicopter primary flight controls. This is illustrated in Figure 7.13, for the fighter helicopter Airbus Tiger.

Figure 7.13. *Moving-rod, active/active tandem, hydro-mechanical main rotor actuator of the Tiger helicopter*

7.5.2. *Structural integration*

The type of structural integration for actuators has to be selected carefully. Indeed, it has a direct influence on forces applied to the airframe (on their nature, intensity and direction), on the geometrical housing dedicated to the actuator as well as on signal and power connections.

The choice of the type of implemented integration has a direct impact on several different aspects: achievable range of motion, forces the airframe has to be able to withstand in reaction to actuation forces, geometrical envelope of the actuator and signal and power connections. In the vast majority of applications, even though the load must be driven rotationally, the actuator consists of a linear hydraulic cylinder. From a kinematic point of view, the hydraulic cylinder then acts as a cylindrical pair joint, whose axial movement is controlled.

7.5.2.1. *Range of motion confined to a few degrees*

When range of motion is restricted to a few dozen degrees, as is the case for primary flight controls, it is possible to directly connect the moving part of the actuator to the driven load. In this situation, translational movement is converted into rotational movement by the load itself thanks to the lever arm effect it generates. This lever arm depends on the position of the actuator anchoring point with respect to the load/airframe pin point. Four different integration alternatives can then be identified [VAL 12]:

– The pin-to-pin mount, depicted in Figure 7.14(a). This integration technique forms a three-bar system, one of which (the one associated with the actuator) has a variable length. This solution is attractive because it is naturally isostatic. However, it comprises two internal degrees of freedom (the cylinder and the rod rotate about their longitudinal axis). Strictly speaking, in order to isostatically implement such integration, connections of the actuator to the airframe and the load should be gimbal-like. This means these connections should only allow rotations about the two axis perpendicular to the cylinder rod axis. In practice, these internal movements are restricted by fitting side pads on spherical bearings. In some cases, guide fingers (body anti-rotation devices) are even added to avoid blocking the load in the event of mechanical rupture at the actuator.

– The trunnion mount, depicted in Figure 7.14(b). This approach is an alternative version of the previous integration method. Here, the connection between the actuator and the airframe takes the form of a cylindrical pair joint. However, in order to guarantee isostatic conditions, it is necessary to ensure a degree of axial freedom. This is achieved by introducing a gap to act as a sliding pin either at the level of trunnions or at the level of the load connecting spherical bearing. This solution is very widely used for spoiler cylinders because it makes it possible to minimize the required longitudinal geometrical envelope.

– The reaction/kick link mount, depicted in Figure 7.14(c). This type of mount locally closes the kinematic structure so as to limit reaction forces the airframe has to withstand. Thanks to the reaction link, the actuator bears on the airframe, as close as possible to its connection with the load, in order to develop actuation forces. As can be seen in Figure 7.15, this solution therefore offers the benefit of reducing the force transmission path length. In turn, this allows the mass of the airframe to be lowered. However, this comes at the cost of three extra hinge-joints, compared with the first solution.

– The flange mount, depicted in Figure 7.14(d). This approach consists of rigidly connecting the actuator body to the airframe. Here, transmission to the load must imperatively limit radial loads and root bending moments absorbed by the rod/cylinder connection of the actuator. This condition is easy to fulfill for the tail rotor pitch controls of helicopters, for which this solution is therefore widespread. This solution can also be used for the control surfaces of aircrafts by introducing a transmission rod (dog bone) if transverse forces seen by the cylinder rod are acceptable.

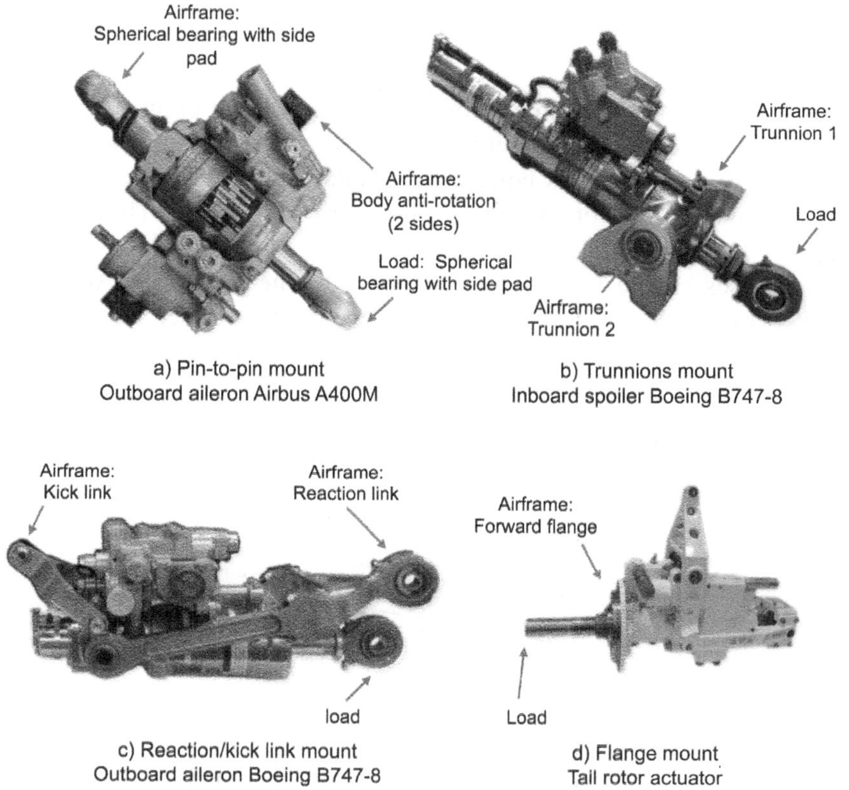

Figure 7.14. *Various types of actuator/airframe integration*

In the first three solutions, the actuator body travels relative to the airframe. This increases the size of the geometrical envelope dedicated to the actuation functions. It also imposes the use of flexible connections at the signal and the power levels (such as hydraulic hoses) which are subjected to fatigue stresses.

7.5.2.2. *Range of motion greater than 90°*

When the total range of motion needed exceeds 90°, as is the case for steering auxiliary landing gear, it is no longer possible to use the simple lever arm effect generated by the load. Indeed, the transmission rate decreases and is cancelled out for a total range of motion of 180°. One possible solution to this issue consists of using a rotary actuator (vane or gear motor for example). However, for landing gear steering purposes, a linear actuator is most often preferred. Its integration can follow one of two structural patterns:

– Rack and pinion movement conversion, depicted in Figure 7.16. From a hydraulic point of view, the actuator acts as a symmetrical linear cylinder that comprises two pistons connected together by a rack. This rack gears with the landing gear rotating tube, which acts as a pinion. This solution is compact. However, the pinion-rack pressure angle develops substantial radial stresses on the rack and the pinion. For the rack, these stresses are absorbed by a roller. This roller is also responsible for setting the angular position of the rack within the cylinder in order to ensure proper gearing.

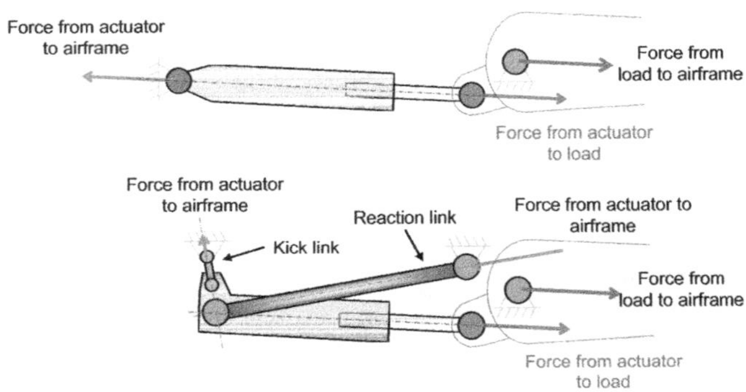

Figure 7.15. *Benefits of the reaction/kick link design on airframe mechanical loading*

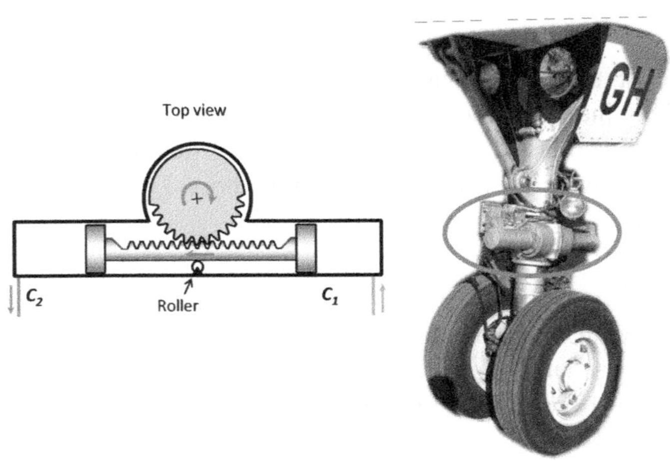

Figure 7.16. *Rack and pinion kinematics for landing gear steering (top view)*

– Push–pull with torque summing movement conversion, depicted in Figure 7.17. In this solution, two linear cylinders are connected to the airframe (the landing gear leg for example) by trunnions of axis parallel to the rotational axis of the load (the landing gear turning tube for example). The torque applied by the actuator on the load is given by the sum of torques developed by each of the two cylinders. The trunnion mount allows a rotating range greater than ±70°. Beyond a certain steering angle, the power supply of one of the cylinders must be reversed (the left one on the example shown in Figure 7.17) in order to switch from case 1 to case 3. This direction commutation is performed mechanically by using the trunnion as a rotaty valve piloted by the relative angle between the cylinder body and the airframe. In practice, the commutation angle is selected in such a way that the take-off and landing phases only call for case 1 (steering angle needed in these phases is of the order of a few degrees, typically 6°). Therefore, commutation is only carried out during taxiing phases requiring large turning radii. From the standpoint of power metering and management, i.e. at ports C_1 and C_2, the assembly consisting of the two cylinders and their valve integrated on trunnions, is equivalent to an asymmetrical double-effect cylinder. Moreover, the lever arm and the active hydrostatic cross-sections vary with the rotation. As a consequence, the overall hydromechanical power transformation ratio also varies with the steering angle. Figure 7.18 illustrates this type of kinematics for Boeing B787 and Airbus A350 WXB.

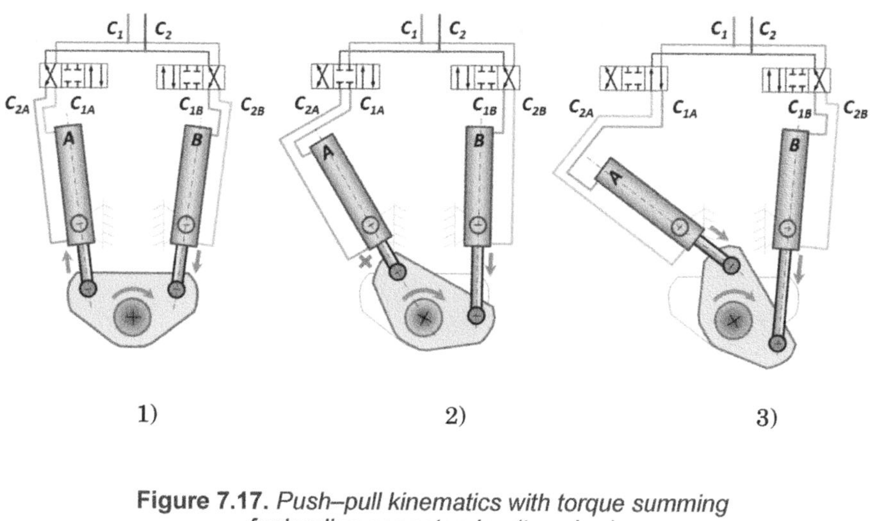

Figure 7.17. *Push–pull kinematics with torque summing for landing gear steering (top view)*

Figure 7.18. *Nose landing gear steering actuator: left: Boeing B787, right: Airbus A350*

7.5.2.3. *Secondary flight controls*

Actuating secondary flight controls (slats, flaps, trim horizontal stabilizer) require the extensive use of mechanical power transmission elements (torque or speed summing, irreversors, brakes, torque limiting devices, etc.). These mechanical power transmission functions will be addressed in the following volume.

Bibliography

[A4A 14] AIR TRANSPORT ASSOCIATION OF AMERICA, iSpec 2200 Extract: ATA Standard Numbering System, January 2014.

[ACC 01] ALL ENGINES-OUT LANDING DUE TO FUEL EXHAUSTION, Accident Investigation Final Report 22/ACCID/GPIAA/2001, 2001.

[AIG 01] AIGAT J., A380 Program Hydraulic Generation and Distribution Equipments, presented to SAE-A6 Committee, Toulouse, 16 October 2012.

[AIR 14] COMMERCIAL AVIATION ACCIDENTS 1958–2013 – A Statistical Analysis, Safety communication Airbus, 2014.

[ATT 08] ATTAR B., Modélisation réaliste en conditions extrêmes des servovalves électrohydrauliques utilisées pour le guidage et la navigation aéronautique et spatiale, Doctoral Thesis, Institut National des Sciences Appliquées de Toulouse, 2008.

[ATC 57] ATCHLEY R.D., Servo-mechanism, US patent 681.310.23, 30 August 1957.

[AUS 83] AUSTIN V.J., "B-1B hydraulics: practical, innovative designs", *Hydraulics and Pneumatics*, pp. 61–66, September 1983.

[BAR 09] ROSENBERG B., "Parker develops a number of hydraulic firsts", *SAE Aerospace Engineering Magazine* (electronic), 18 August 2009.

[BEA 08] BUREAU ENQUETE ET ANALYSES, Incidents en transport aérien – Pannes hydrauliques, no. 9, p. 1, June 2008.

[BEH 10] BEHR R., BAUMBACH V., "Model Based Aerospace Hydraulic Fluid Lifetime Prognostics", *Proceedings of the 4th International Conference on Recent Advances in Aerospace Actuation Systems and Components*, Toulouse, pp. 168–173, May 5–7, 2010.

[BEH 13] BEHR R., BAUMBACH V., "Aircraft hydraulic thermal control", *Proceedings of Aircraft Systems Technology Workshop*, Hamburg, Germany, April 23–24, 2013.

[BEN 93] BENNETT S., "The PID controller: the story of an idea", *Proceedings of the 2nd European Control Conference*, Groningen, Netherlands, pp. 1235–1239, June 28–July 1, 1993.

[BIE 98] BIEDERMANN O., GEERLING G.,"Power control units with secondary controlled hydraulic motors – a new concept for application in aircraft high lift systems", *Proceedings of the International Conference on Recent Advances in Aerospace Hydraulics*, Toulouse, pp. 73–77, November 24–25, 1998.

[BLA 60] BLACKBURN J.F., REETHOF G., SHEARER J.L., *Fluid Power Control*, Wiley, New-York, February 15, 1960.

[BOE 15] STATISTICAL SUMMARY OF COMMERCIAL JET AIRPLANE ACCIDENTS – WORLDWIDE OPERATIONS 1959–2014, Aviation Safety – Boeing Commercial airplanes, August 2015.

[BOW 04] BOWERS J., "A380 Hydraulic Slat Drive Channel", *Proceedings of the International Conference on Recent Advances in Aerospace Actuation Systems and Components*, Toulouse, France, pp. 91–96, November 24–26, 2004.

[BOY 55] BOYAR R.E., JOHNSON B.A., SCHMIDT L., "Hydraulic servo control valve–Part 1: a summary of the present state of the art of electrohydraulic servovalves", *WADC technical report 55–29 part 1*, Wright Patterson air force base, pp. 23–25, Ohio, 1955.

[BUR 82] BURNETT J., GOLDMAN P., MC ADAMS F. *et al.*, "Eastern Airlines Flight 935, Lockheed L-1011-385, N309EA, Near Colts Neck, New Jersey, September 22, 1981", *Official Incident Report* No. *NTSB-AAR-82-5*, pp. 1–61, June 1, 1982.

[CAL 04] CALLENARE M., SAUTERY G., "Fluid velocity limits in hydraulic systems: detrimental effects on distribution systems", *Proceedings of the 2nd International Conference on Recent Advances in Aerospace Actuation Systems and Components.* Toulouse, pp. 141–146, November 24–26, 2004.

[CAR 53] CARSON H., Flow control servovalve, US patent 2.934.765, September 21, 1953.

[CCP 05] CONCOURS COMMUNS POLYTECHNIQUE, Etude du Plan Horizontal Réglable (PHR) de l'AIRBUS A340, Sujet de l'épreuve spécifique de Sciences Industrielles – filière MP, 2005.

[CLA 57] CLARK R.N., "Compensation of steady-state flow forces in spool-type hydraulic valves", *ASME Trans.*, pp. 1784–1788, November 1957.

[CND 02] CENTRE NATIONAL DE DOCUMENTATION PEDAGOGIQUE, Initiation aux Sciences de l'Ingénieur, October 2002.

[COM 05] COMES M., L'A380, Nouveautés techniques ou techniques nouvelles, *Conférence Association Aéronautique et Astronautique de France*, Toulouse, 7 January 2005.

[COW 72] COWIN H., Monofuel emergency power systems, Flight International, p. 607, 2 November 1972.

[CRA 99] CRANE D., *Aircraft Hydraulic Systems*, Aviation Supply and Academics, 1999.

[CUR 88] CURREY N.S., *Aircraft Landing Gear Design: Principles and Practices*, American Institute of Aeronautics & Astronautics Education series, June 1988.

[DAC 04] DA COSTA A., VILLENEUVE M., "Recent Developments in High Pressure Maintenance Free Accumulators for 5000 PSI Aircraft Hydraulics", *Proceedings of the 2nd International Conference on Recent Advances in Aerospace Actuation Systems and Components.* Toulouse, pp. 147–152, November 24–26, 2004.

[DAN 15] DANIEL G., TAGLIAVINI A., MOSCA P., *Ouvrage 21-A: Connaissance générale des aéronefs – tome 1*, Editions Jean Mermoz, 2015.

[DAV 75] DAVIS H.D., "Shuttle's orbiter: giant hydraulic step in space", *Hydraulics and Pneumatics*, pp. 45–48, December 1975.

[DEL 04] DELLAC S., TERNISIEN D., "Airbus 380 "Electro-Hydraulic Back-Up Architecture" for Braking and Steering Systems", *Proceedings of the 2nd International Conference on Recent Advances in Aerospace Actuation Systems and Components*, Toulouse, pp. 103–108, November 24–26, 2004, France.

[DIE 14] SCHOLZ D., "Aircraft Cabin Air & Water Contamination/Quality – An Aircraft Systems Engineering Perspective", *Global Cabin Air Quality Executive – 7th Annual Forum and Information Exchange*, London, 31 March–2 April, 2014.

[DON 40] MCDONOUGH E.W., *Aircraft Hydraulic Systems – A Text on Applied Fluid Mechanics*, Aircraft Specialties Co. Chicago, 1940.

[EAT 00] EATON DOCUMENTATION, A Descriptive Summary of Vickers Inline Pumps and their Applications, Document SE-103F, June 2000.

[EAT 04] EATON CORPORATION AEROSPACE, Airbus A380, Eatons's Aerospace product capability C5-37, February 2004.

[EDU 13] EDUSCOL, Portail National des Professionnels de l'Education, Dossier ressource de la Ram Air Turbine, available at: http://eduscol.education.fr, 2013.

[FAA 01] FEDERAL AVIATION ADMINISTRATION, FAA 14 CFR Part 25.1435 – Airworthiness standards: transport category airplanes – Hydraulic systems, 16 May 2001.

[FAA 12] FEDERAL AVIATION ADMINISTRATION, Aviation Maintenance Technician Handbook –Airframe, FAA-H-8083-31, Available at: https://www.faa.gov, 2012.

[FAA 14] FEDERAL AVIATION ADMINISTRATION, System Safety Handbook – Appendix A: Glossary, Create Space Independent Publishing Platform, 30 July 2014.

[FAI 81] FAISANDIER J., *Hydraulique et électrohydraulique*, Dunod, Paris, 1981.

[FAI 99] FAISANDIER J. et coll., *Mécanismes hydrauliques et pneumatiques*, Dunod, 1999.

[FAY 91] FAYET G., *Hydraulique – Machines et composants*, Eyrolles, Paris 1991.

[FEN 05] FENNY C.A., SCHULTZ D.P., "Design and development of the BA609 civil tiltrotor hydraulic architecture", *American Helicopter Society 61st Annual Forum*, Grapevine, Texas, June 1–3, 2005.

[GEY 72] GEYER L.H., "Controlled damping through dynamic pressure feedback", *Moog technical bulletin 101*, April 1972.

[GRE 85] GREEN W.L., *Aircraft Hydraulic Systems*, John Wiley, 1985.

[GUI 72] GUILLON M., *L'asservissement hydraulique et électrohydraulique*, Dunod, Paris, 1972.

[GUI 92] GUILLON M., *Commande et asservissement hydrauliques et électrohydrauliques*, Tec et doc, Lavoisier, 1992.

[HAL 98] HALAT J.A., GALLOWAY P.W., "High pressure hydraulic pumps", *Proceedings of the Internaional Conference on Recent Advances in Aerospace Hydraulics*, Toulouse, France, pp. 25–30, November 24–25, 1998.

[HSU 91] HSU Y.-N., LAI C.-Y., HSU M.-H. *et al.*, "Development of the DDV actuation system on the IDF aircraft", SAE paper 912080, *Air and Space Technology Conference and 29th Aircraft Symposium Proceedings*, pp. 871–881, Japan, 7–11 October 1991.

[ICA 93] INTERNATIONAL CIVIL AVIATION ORGANIZATION, "Manual of the ICAO Standard Atmosphere (extended to 80 kilometers (262 500 feet)", *Standard of the International Civil Aviation Organization*, Doc 7488-CD, 3rd ed., 1993.

[ISE 06] ISERMANN R., *Fault Diagnosis Systems*, Springer, Berlin, 2006.

[IVA 01] IVANTYSYN J., IVANTYSYNOVA M., *Hydrostatic Pumps and Motors*, Academia Books International, New-Delhi, 2001.

[JAC 54] Servocommandes de vol Jaccottet-Leduc, Documentation technique Jaccottet-Leduc, Courbevoie, 1954.

[JEF 98] JEFFERY J., "A brief history of high pressure hydraulics", *Proceedings of the International Conference on Recent Advances in Aerospace Hydraulics,* Toulouse, pp. 31–61, November 24–26, 1998.

[JEP 85] JEPPESEN SANDERSON, *Aircraft Hydraulic Systems*, 1985.

[JOH 05] JOHANSSON A., Design principles for noise reduction in hydraulic piston pumps: simulation, optimisation and experimental verification, Doctoral Thesis, Linköping University, 2005.

[KEL 74] KELLER G.R., *Hydraulic System Analysis*, Academia Books International, New-Delhi, Penton/IPC, 1974.

[KAU 10] KAUSCH M., MEYER M., KROLL L. *et al.,* "Novel hydraulic lightweight manifolds in aviation", *Proceedings of the 4th International Conference on Recent Advances in Aerospace Actuation Systems and Components*, Toulouse, France, pp. 56–59, May 5–7, 2010.

[LIN 86] LINTON D.J., "Energy efficient actuation using variable displacement hydraulic control", *Advanced Actuation Controls and Integration for Aerospace Vehicles*, P-170, Paper 851755, Society of Automotive Engineers, Warrendale, Pennsylvania, pp. 51–55, 1996.

[LIS 08] LISCOUËT-HANKE S., A model-based methodology for integrated preliminary sizing and analysis of aircraft power system architecture, Doctoral Thesis, Institut national des Sciences Appliquées de Toulouse, October 2008.

[LIS 09] LISCOUËT-HANKE S., MARÉ J.-C., PUFE S., "Simulation Framework for Aircraft Power System Architecting", *Journal of Aircraft*, vol. 46, no. 4, pp. 1375–1380, 2009.

[MAD 84] MADISSON R.T., "The MD-80: hydraulics simple, reliable, and easy to operate", *Hydraulics and Pneumatics*, 5 July 1984.

[MAL 04] MALINGE Y., "A300B4 Loss of All Hydraulics. Baghdad: A Remarkable Example of Airmanship", *Proceedings of the International Society of Air Safety Investigator*, Queensland, Australia, pp. 87–89, August 30–September 2, 2004.

[MAR 80] MARTIN H.R., Mc Cloy D., *Control of Fluid Power*, Ellis Horwood, 1980.

[MAR 02a] MARÉ J.-C., “Actionneurs hydrauliques – Conception préliminaire”, Article S7530, série Mesure et Régulation, Techniques de l’Ingénieur, September 2002.

[MAR 02b] MARÉ J.-C., “Actionneurs hydrauliques – Commande”, Article S7531, série Mesure et Régulation, Techniques de l’Ingénieur, September 2002.

[MAR 04] MARÉ J.-C., “Les innovations de l’A380”, *Revue Fluides, Numéro Spécial Matériels*, pp. 54–59, September 2004.

[MAR 09] MARÉ J.-C., BUDINGER M., “Comparative analysis of energy losses in servo-hydraulic, electro-hydrostatic and electro-mechanical actuators”, *Proceedings of the 11th Scandinavian International Conference on Fluid Power*, Linköping, pp. 66–67, June 2–4, 2009.

[MAR 10] MARGER T., Conception d’un distributeur de servocommande hydromécanique sous critères de coût et de mixabilité, Doctoral Thesis, Arts et Métiers ParisTech– Centre d’Aix-en-Provence, 13 December 2010.

[MAR 11] MARÉ J.-C.,”Combining hydraulics and electrics for innovation and performance improvement in aerospace actuation”, *Proceedings of the 12th Scandinavian International Conference on Fluid Power*, Tampere, May 18–20, 2011.

[MAS 78] MASKREY D.H., THAYER W.J., “A brief history of electrohydraulic servomechanisms”, *Journal of Dynamic Systems Measurement and Control*, vol. 11,no. 2, pp. 110–116, 01 June 1978.

[MCF 67] MC FADDEN E.F., WILLIAMS L., Free jet stream deflector servovalve, US patent 694.739, 29 December 1967.

[MEI 98] MEINADIER J.-P., *Ingénierie et intégration des systèmes*, Hermes Science, Paris, 1998.

[MER 67] MERRITT H.E., *Hydraulic Control Systems*, Wiley, 1967.

[MIL 86] MILITARY SPECIFICATION: Hydraulic fluid, fire resistant, synthetic hydrocarbon base MIL-H-83282C, 25 March 1986.

[MIL 93] MILLER F.G., “Direct drive control valves and their applications”, *Proceedings of the Aerospace Hydraulics and Systems International Conference, IMECH 1993-6*, London, pp. 1–16, 29–30 September 1993.

[MOI 01] MOIR I., SEABRIDGE A., *Aircraft Systems: Mechanical, Electrical, and Avionics Subsystems Integration*, 3rd ed., John Wiley & Sons, 2008.

[NEE 91] NEESE W., *Aircraft Hydraulic Systems*, 3rd ed., Krieger Publishing Company, Malabar, 1991.

[NGU 04] NGUYEN T., DE LA CHEVASNERIE A., SANDLER S., "Direct drive servo valve", *Proceedings of the 2nd International Conference on Recent Advances in Aerospace Hydraulics*, Toulouse, France, pp. 73–78, November 24–26, 2004.

[OCO 12] O'CONNOR P., KLEYNER A., *Practical Reliability Engineering*, Wiley–Blackwell, 6 January 2012.

[RAY 93] RAYMOND E.T., CHENOWETH C.C., *Aircraft Flight Control Actuation System Design*, SAE Press, 1993.

[ROS 00] ROSKAM J., *Airplane Design Part IV – Layout of Landing Gear and Systems*, Design and Research corporation, 3rd print, 2000.

[SAE 04] SOCIETY OF AUTOMOTIVE ENGINEERS, 8000 psi hydraulic systems: experience and test results, Aerospace Information Report 4002, 18 March 2004.

[SAE 08] SOCIETY OF AUTOMOTIVE ENGINEERS, Aerospace Hydraulic Fluids Physical Properties, Standard SAE 1362-B, 17 July 2008.

[SAE 97] SOCIETY OF AUTOMOTIVE ENGINEERS, Fire Resistant Phosphate Ester Hydraulic Fluid for Aircraft, Standard SAE 1241-C, 1 September 1997.

[SAE 11] SOCIETY OF AUTOMOTIVE ENGINEERS, "Graphic symbols for aircraft hydraulic and pneumatic systems", *Aerospace Standard AS 1290-B*, 12 December 2011.

[SAN 07] SANDLER S., HERVIEUX A., "Electrical Flight Control Technologies for Rotorcrafts", *Proceedings of the 3rd International Conference on Recent Advances in Aerospace Actuation Systems and Components*, Toulouse, France, pp. 9–14, June 13–15, 2007.

[SAU 09] SAUTERET D., *Ouvrage 20: Compléments hélicoptère CPL*, Editions Jean Mermoz, 2009.

[SCH 98] SCHWEITZER J.J., BARROW G., "Reliability and service experience of aerospace hydraulic rotating equipment", *Proceedings of the International Conference on Recent Advances in Aerospace Hydraulics*, Toulouse, France, pp. 1–6, November 24–26, 1998.

[SCH 93] SCHAEFFER W.S., INDERHEES L.J., MOYNES J.F., "Flight control actuation for the B-2 advanced technology bomber", *Proceedings of the Aerospace Hydraulics and Systems international conference,* IMECH 1993-6, London, pp. 23–32, 29–30 September 1993.

[SCH 98] SCHMITT R.S., MORRIS J.R., *Fly by Wire*, SAE Press, 1998.

[SEW 94] SEWARD D., "120 Killed In Russian Air Crash; Safety Concerns Rise in Ex-U.S.S.R.", *The Washington Post*, 4 January 1994.

[SWE 51] SWEENEY D.C., "Preliminary investigation of hydraulic lock", *Engineering*, pp. 513–516 and 580–582, vol. 72, 1951.

[TAK 87] TAKEDA S., ENOMOTO Y., NISHIMURA K. *et al.*, Aircraft Accident Investigation Report, Japan Air Lines – Boeing 747 SR-100 – JA8119 – Japan – August 12, 1985, pp. 1–333, June 19, 1987.

[TAY 65] THAYER W.J., "Transfer function for Moog servovalves", *Moog technical bulletin 103*, January 1965.

[TAY 76] TAYLER W.J., "Redundant electrohydraulic servoactuators", *Moog technical bulletin 127*, pp. 1–32, May 1976.

[TAY 80] TAYLER W.J., "Understanding redundant electrohydraulic servoactuators", *Hydraulics & Pneumatics*, pp. 22–66, June 1980.

[TCH 79] TCHOUPRAKOV Y., *Commande hydraulique et automatismes hydrauliques*, Editions MIR, Moscow, 1979.

[THO 42] THOMPSON J.E., CAMPBELL R.B., *Manual for Aircraft Hydraulics, Theory, Maintenance, Design*, Aviation Press, San Francisco, 1942.

[TIN 49] TINSLEY H.V., English patent 620.688, May 1946.

[TRA 06] TRAVERSE P., LACAZE I., SOUYRIS J., "Airbus Fly-by-wire: a process toward total dependability", *Electronic Proceedings of the ICAS 2006 25th International Congress of the Aeronautical Sciences*, Hamburg, Germany, 3–8 September 2006.

[VAL 12] VALDO M., Servo-hydraulic technology in flight control, *Workshop on Innovative Engineering for Fluid Power and Vehicular Systems*, ABIMAQ – São Paulo, pp. 1–72, 14–15 May 2012.

[VAN 90] VAN BROCKLIN C., "Information compendium on nonflammable hydraulic fluid and design requirements for its adoption", *Report 2486, AD-A218 569*, US Army, Fort Belvoir, February 1990.

[VIC 98] Vickers product news, Constant pressure reversible power transfer unit, document 123-1, Mars 1998.

[VIE 80] VIERSMA T.J., *Analysis, Synthesis and Design of Hydraulic Servosystems and Pipelines*, Elsevier, 1980.

[WIL 08] WILD T.W., *Transport Category Aircraft Systems*, 3rd ed., Jeppesen, 2008.

[WIL 01] WILKINSON R., *Aircraft Structures and Systems*, 2nd ed., Mechaaero Publishing, 2001.

[WIL 50] WILLIAM C., MOOG J.R., Electrohydraulic servomechanism, US patent 2.625.136, 26 April 1950.

[WIL 53] WILLIAM C., MOOG J.R., Electrohydraulic servovalve, US patent 2.767.689, 22 May 1953.

[WUN 38] WUNSCH G., HARTUNG W., Jet pipe relay, US patent US 2.107.971 A, 8 February 1938.

Index

A, B, C

D, E

F

G, H, I

J, L, M

N, O, P

R, S

T, U

Other titles from

iSTE

in

Systems and Industrial Engineering – Robotics

2016

AUBRY Jean-François, BRINZEI Nicolae, MAZOUNI Mohammed-Habib
Systems Dependability Assessment: Benefits of Petri Net Models

2015

AUBRY Jean-François, BRINZEI Nicolae
Systems Dependability Assessment: Modeling with Graphs and Finite State Automata

BOULANGER Jean-Louis
CENELEC 50128 and IEC 62279 Standards

BRIFFAUT Jean-Pierre
E-Enabled Operations Management

MISSIKOFF Michele, CANDUCCI Massimo, MAIDEN Neil
Enterprise Innovation

2014

CHETTO Maryline
Real-time Systems Scheduling
Volume 1 – Fundamentals
Volume 2 – Focuses

DAVIM J. Paulo
Machinability of Advanced Materials

ESTAMPE Dominique
Supply Chain Performance and Evaluation Models

FAVRE Bernard
Introduction to Sustainable Transports

GAUTHIER Michaël, ANDREFF Nicolas, DOMBRE Etienne
Intracorporeal Robotics: From Milliscale to Nanoscale

MICOUIN Patrice
Model Based Systems Engineering: Fundamentals and Methods

MILLOT Patrick
Designing Human–Machine Cooperation Systems

NI Zhenjiang, PACORET Céline, BENOSMAN Ryad, REGNIER Stéphane
Haptic Feedback Teleoperation of Optical Tweezers

OUSTALOUP Alain
Diversity and Non-integer Differentiation for System Dynamics

REZG Nidhal, DELLAGI Sofien, KHATAD Abdelhakim
Joint Optimization of Maintenance and Production Policies

STEFANOIU Dan, BORNE Pierre, POPESCU Dumitru, FILIP Florin Gh., EL KAMEL Abdelkader
Optimization in Engineering Sciences: Metaheuristics, Stochastic Methods andDecision Support

2013

ALAZARD Daniel
Reverse Engineering in Control Design

ARIOUI Hichem, NEHAOUA Lamri
Driving Simulation

CHADLI Mohammed, COPPIER Hervé
Command-control for Real-time Systems

DAAFOUZ Jamal, TARBOURIECH Sophie, SIGALOTTI Mario
Hybrid Systems with Constraints

FEYEL Philippe
Loop-shaping Robust Control

FLAUS Jean-Marie
Risk Analysis: Socio-technical and Industrial Systems

FRIBOURG Laurent, SOULAT Romain
Control of Switching Systems by Invariance Analysis: Application to Power Electronics

GROSSARD Mathieu, REGNIER Stéphane, CHAILLET Nicolas
Flexible Robotics: Applications to Multiscale Manipulations

GRUNN Emmanuel, PHAM Anh Tuan
Modeling of Complex Systems: Application to Aeronautical Dynamics

HABIB Maki K., DAVIM J. Paulo
Interdisciplinary Mechatronics: Engineering Science and Research Development

HAMMADI Slim, KSOURI Mekki
Multimodal Transport Systems

JARBOUI Bassem, SIARRY Patrick, TEGHEM Jacques
Metaheuristics for Production Scheduling

KIRILLOV Oleg N., PELINOVSKY Dmitry E.
Nonlinear Physical Systems

LE Vu Tuan Hieu, STOICA Cristina, ALAMO Teodoro, CAMACHO Eduardo F., DUMUR Didier
Zonotopes: From Guaranteed State-estimation to Control

MACHADO Carolina, DAVIM J. Paulo
Management and Engineering Innovation

MORANA Joëlle
Sustainable Supply Chain Management

SANDOU Guillaume
Metaheuristic Optimization for the Design of Automatic Control Laws

STOICAN Florin, OLARU Sorin
Set-theoretic Fault Detection in Multisensor Systems

2012

AÏT-KADI Daoud, CHOUINARD Marc, MARCOTTE Suzanne, RIOPEL Diane
Sustainable Reverse Logistics Network: Engineering and Management

BORNE Pierre, POPESCU Dumitru, FILIP Florin G., STEFANOIU Dan
Optimization in Engineering Sciences: Exact Methods

CHADLI Mohammed, BORNE Pierre
Multiple Models Approach in Automation: Takagi-Sugeno Fuzzy Systems

DAVIM J.Paulo
Lasers in Manufacturing

DECLERCK Philippe
Discrete Event Systems in Dioid Algebra and Conventional Algebra

DOUMIATI Moustapha, CHARARA Ali, VICTORINO Alessandro, LECHNER Daniel
Vehicle Dynamics Estimation using Kalman Filtering: Experimental Validation

GUERRERO José A, LOZANO Rogelio
Flight Formation Control

HAMMADI Slim, KSOURI Mekki
Advanced Mobility and Transport Engineering

MAILLARD Pierre
Competitive Quality Strategies

MATTA Nada, VANDENBOOMGAERDE Yves, ARLAT Jean
Supervision and Safety of Complex Systems

POLER Raul *et al.*
Intelligent Non-hierarchical Manufacturing Networks

TROCCAZ Jocelyne
Medical Robotics

YALAOUI Alice, CHEHADE Hicham, YALAOUI Farouk, AMODEO Lionel
Optimization of Logistics

ZELM Martin *et al.*
Enterprise Interoperability –I-EASA12 Proceedings

2011

CANTOT Pascal, LUZEAUX Dominique
Simulation and Modeling of Systems of Systems

DAVIM J. Paulo
Mechatronics

DAVIM J. Paulo
Wood Machining

GROUS Ammar
Applied Metrology for Manufacturing Engineering

KOLSKI Christophe
Human–Computer Interactions in Transport

LUZEAUX Dominique, RUAULT Jean-René, WIPPLER Jean-Luc
Complex Systems and Systems of Systems Engineering

ZELM Martin, *et al.*
Enterprise Interoperability: IWEI2011 Proceedings

2010

BOTTA-GENOULAZ Valérie, CAMPAGNE Jean-Pierre, LLERENA Daniel, PELLEGRIN Claude
Supply Chain Performance / Collaboration, Alignement and Coordination

BOURLÈS Henri, GODFREY K.C. Kwan
Linear Systems

BOURRIERES Jean-Paul
Proceedings of CEISIE'09

CHAILLET Nicolas, REGNIER Stéphane
Microrobotics for Micromanipulation

DAVIM J. Paulo
Sustainable Manufacturing

GIORDANO Max, MATHIEU Luc, VILLENEUVE François
Product Life-Cycle Management / Geometric Variations

LOZANO Rogelio
Unmanned Aerial Vehicles / Embedded Control

LUZEAUX Dominique, RUAULT Jean-René
Systems of Systems

VILLENEUVE François, MATHIEU Luc
Geometric Tolerancing of Products

2009

DIAZ Michel
Petri Nets / Fundamental Models, Verification and Applications

OZEL Tugrul, DAVIM J. Paulo
Intelligent Machining

PITRAT Jacques
Artificial Beings

2008

ARTIGUES Christian, DEMASSEY Sophie, NERON Emmanuel
Resources–Constrained Project Scheduling

BILLAUT Jean-Charles, MOUKRIM Aziz, SANLAVILLE Eric
Flexibility and Robustness in Scheduling

DOCHAIN Denis
Bioprocess Control

LOPEZ Pierre, ROUBELLAT François
Production Scheduling

THIERRY Caroline, THOMAS André, BEL Gérard
Supply Chain Simulation and Management

2007

DE LARMINAT Philippe
Analysis and Control of Linear Systems

DOMBRE Etienne, KHALIL Wisama
Robot Manipulators

LAMNABHI Françoise *et al.*
Taming Heterogeneity and Complexity of Embedded Control

LIMNIOS Nikolaos
Fault Trees

2006

FRENCH COLLEGE OF METROLOGY
Metrology in Industry

NAJIM Kaddour
Control of Continuous Linear Systems

Printed and bound by CPI Group (UK) Ltd, Croydon, CR0 4YY
13/09/2026
14961021-0003